Water Governance—Concepts, Methods, and Practice

Series editors

Claudia Pahl-Wostl, University of Osnabrück, Osnabrück, Germany
Joyeeta Gupta, Faculty of Earth and Life Sciences, University of Amsterdam, Amsterdam, The Netherlands

This book series aims at providing a platform for developing an integrated perspective on major advances in the field of water governance. Contributions will build bridges across established fields of expertise, across disciplinary perspectives, across levels from global to local and across geographical regions. Topics to be covered include conceptual and methodological advances capturing the complexity of water governance systems, institutional settings, actor constellations, diagnostic approaches, and comparative studies of governance systems. The book series encompasses monographs, textbooks and edited, coordinated volumes addressing one theme from different perspectives. The book series will address mainly a wider scientific audience, but will also provide valuable knowledge to interested practitioners. The sustainable management of fresh water resources and the ecosystem services that they provide is one of the key challenges of the 21st century. Many water related problems can be attributed to governance failure at multiple levels of governance rather than to the resource base itself. At the same time our knowledge on water governance systems and conditions for success of water governance reform is still quite limited. The notion of water governance aims at capturing the complexity of processes that determine the delivery of water related services for societal needs and that provide the context within which water management operates. Water governance is a fast growing field of scholarly expertise which has largely developed over the past decade. The number of publications in peer reviewed journals has increased from less than 20 in the year 2000 to nearly 400 in the year 2013. The increasing popularity of the term in science and policy has not lead to conceptual convergence but rather to an increasing vagueness and competing interpretations of how the concept should be understood, studied, and analyzed; which disciplines are involved; and what methodological approaches are most suitable for the study and analysis of water governance. The time seems to be ripe for comprehensive synthesis and integration.

More information about this series at http://www.springer.com/series/13400

Emmy Bergsma

From Flood Safety to Spatial Management

Expert-Policy Interactions in Dutch and US Flood Governance

Springer

Emmy Bergsma
University of Amsterdam
Amsterdam, The Netherlands

ISSN 2365-4961 ISSN 2365-497X (electronic)
Water Governance—Concepts, Methods, and Practice
ISBN 978-3-030-07236-0 ISBN 978-3-319-96716-5 (eBook)
https://doi.org/10.1007/978-3-319-96716-5

Printed on acid-free paper

This Springer imprint is published by the registered company Springer International Publishing AG part of Springer Nature
The registered company address is: Gewerbestrasse 11, 6330 Cham, Switzerland

Acknowledgements

The research for this manuscript was carried out at the Department of Political Science, Faculty of Social and Behavioural Sciences, the Amsterdam Institute for Social Science Research (AISSR), University of Amsterdam. It was made possible by the *Stichting Agnosticisme and Meritocratie*.

Contents

Abbreviations

FEMA	Federal Emergency Management Agency
FIRM	Flood Insurance Rate Map
GAO	US Government Accountability Office
NFIP	National Flood Insurance Program
US	United States

Chapter 1
A Framework for Analyzing Distributive Decision-Making in Flood Governance

1.1 Problem: Who Pays for Floods?

In the last decades, the costs of floods have risen. In 2017 alone, more than 1200 people were killed during monsoon floods in South Asia, hurricanes Harvey and Irma brought 200–300 billion US dollar (USD) damage to the Gulf Coast and the Caribbean, and large parts of central Europe were hit by massive riverine flooding. 2017 is no exception. For years, floods have ranked on top of the United Nations Office for Disaster Risk Reduction's list of most disastrous natural hazards (UNISDR 2015). In a future of climate change and continued urbanization of delta regions, the impacts of flood events are only expected to increase (Bouwer 2010: 105; Swiss Re 2012: 6; World Economic Forum 2015). Experts have projected the number of people affected by riverine floods to increase from 21 to 54 million by 2030 (Winsemius et al. 2013) while the global exposure to river and coastal flooding may rise from 42 to 180 trillion USD by 2050 (Jongman et al. 2012), and the poor will be hit hardest (Winsemius et al. 2018). A key question that emerges from these rising (future) costs of flooding is who will pay for these costs and the protection measures to avert them? This question forms the starting point of this book.

1.1.1 Distributive Choices in Environmental Governance

This study is the result of a personal quest into an issue that has bothered me since the start of my studies in human geography: Who pays for natural distress? Human geography analyses the relationship between humans and their (social and natural) environment; my specialization, environmental geography, focuses on our relationship with nature specifically. This specialization looks at how we make use of our natural environment, and how we create the conditions that allow us to access natural areas, utilize their resources, and maximize their benefits. It also looks at the impli-

E. Bergsma, *From Flood Safety to Spatial Management*, Water Governance - Concepts, Methods, and Practice, https://doi.org/10.1007/978-3-319-96716-5_1

cations of these human interferences with nature. By shaping nature to our needs, we have exposed ourselves to its harms. We have not only contributed to environmental problems such as earthquakes, biodiversity loss, droughts, and floods but these problems now also present a threat to the environment we have become so dependent on (Goudie 2000). In the management of environmental problems, the short-term and often individual benefits gained by our use of nature have to be considered against the long-term and often collective risks created by this use. How to make such trade-offs is, in my view, one of the most pressing and relevant questions in environmental governance.

Over the years, we have become increasingly aware of humans' role in creating environmental hazards. The 1987 Brundtland *Our Common Future* report already drew attention to our "unsustainable uses" of nature, ideas that have recently been reemphasized under the header of "the Anthropocene". This has changed the way we look at environmental risks. In traditional paradigms, environmental risks were considered an "act of God" or an "extreme" force of nature. They could at best be "reduced to a fair degree of certainty by statistical grouping" by quantifying their probabilities, gaining some form of control over these events (Knight 1921: 215). Collective state-responses were the justified means through which protection was provided. The increased recognition of anthropogenic causes underlying environmental risks has changed this standard policy discourse (Beck 1992; Giddens 1999). When environmental distress can be attributed to human action, collective state-responses lose their forthright appropriateness. A better understanding of anthropogenic contributions to creating environmental problems thus sheds new light on the trade-off between individual benefits and collective risks in environmental risk governance.

1.1.2 *Distributive Choices in Flood Governance: The Shift to Spatial Measures*

Against this background, I grew interested in flood risk management. Flood-prone areas harbor fertile agricultural lands, are often strategically located for international shipping and commerce, and provide good conditions for economic production. Many cities emerged along the shores of river and coastal deltas, and the expectation is that more people will be drawn to these regions in the future. However, these people face serious risks. Delta regions are known for catastrophic flooding disasters, demonstrated for example by the tremendous havoc caused by Hurricane Katrina in the Gulf Coast region in 2005.

Traditionally, societies have always responded to increasing flood risks by building structural flood defenses, such as levees and floodwalls. However, because of the rising costs of flooding disasters in delta regions, questions have emerged about the continued appropriateness of this traditional governance strategy. Can we continue to build our way out of increased flood risks? If so, against what costs? And who will pay for these costs?

Although the importance of traditional safety measures continues to be underlined, the limitations of this approach have been outlined as well (Sayers et al. 2002; Hartmann 2009). Levees and floodwalls foster a feeling of safety that attracts new developments to vulnerable areas. As climate change brings forward more extreme weather conditions, levees will need to be built even higher to continue to protect these settlements. At the same time, levee failures will cause more damage because they protect more people and more value. Under a safety approach, national governments not only have to keep investing in flood protection to continue to avert these costs but because they have involved themselves in flood protection, they are also increasingly held liable for the costs of emergency management and damage compensation.

Against this background, flood-prone countries are demonstrating an increased interest in spatial measures to control flood risks (Pahl-Wostl 2007; Fournier et al. 2016). Spatial measures accommodate floods rather than build against them (Neuvel and Van der Knaap 2010). Their goal is to reduce the impact of a flood by making sure floodwater can enter and leave the physical landscape without causing (much) damage. They for example include the creation of natural overflow areas, the development of land-use (planning) regulations that limit developments in flood-prone areas, and the flood-proofing of buildings in high-risk areas (e.g., the elevation of buildings, the use of water-resistant construction materials, and moving important utilities such as fuse boxes to upper floors). Because spatial measures do not build up the flood risk but reduce the vulnerability to floods, they can be a cost-efficient alternative to structural flood protection (Sayers et al. 2002; Botzen and Van den Bergh 2008; Liao 2012). Therefore, spatial measures are seen as part of a more "risk-based" approach to flood governance, where not only flood chances but also flood impacts are taken into account (De Vries and Wolsink 2009; Merz et al. 2010).

Many flood-prone countries have started to include spatial measures in their portfolio of flood risk governance strategies (Bubeck et al. 2012; Van Buuren et al. 2012; Hegger et al. 2016). Spatial measures can have huge implications for the division of costs and responsibilities in flood governance. In a traditional safety approach to floods, the national government assumes the largest part of the costs of flood protection and damage compensation (Pahl-Wostl et al. 2011). A characteristic of spatial measures, however, is the involvement they require of regional, local, private and individual actors (Johnson and Priest 2008; Merz et al. 2010; Butler and Pidgeon 2011; Bergsma et al. 2012; Rijke et al. 2012; Doorn 2016; Hegger et al. 2017). For example, regulations to limit developments in flood-prone areas demand actions from regional and local authorities charged with spatial planning tasks (Wolsink 2006; Meijerink and Dicke 2008). Reserving space for natural overflow means that the people living in these overflow areas will increasingly suffer periodic flood damage (Penning-Rowsell and Pardoe 2012; Keessen et al. 2016; Paudel et al. 2015).

As of yet, the distributive implications of spatial measures have not received much attention in the academic literature on flood risk management. Of the studies that address this topic, some indeed found costs and responsibilities shift to the regional and local level as a result of the implementation of spatial measures in flood governance (e.g., Butler and Pidgeon 2011; Penning-Rowsell and Pardoe 2012; Keessen et al. 2016; Paudel et al. 2015). Others, however, concluded that the implementation of spatial measures has not yet brought any fundamental changes to traditional responsibility relations in flood governance (Gralepois et al. 2016; Wiering et al. 2017). In these cases, responsibilities and costs for flood protection and damage compensation were still primarily borne by national governments. These divergent findings indicate that under a spatial approach to floods, different distributions of costs and responsibilities are possible (on the diversity underlying spatial approaches, see Aakre et al. 2010; Hegger et al. 2016, 2017; Wiering et al. 2017).

It is important that the distributive implications of a shift to spatial measures are well understood. As countries are starting to implement spatial flood governance policies, policymakers have to decide how to distribute the costs and responsibilities connected to these measures. If regional, local, private and individual actors receive new responsibilities for flood risk protection under a spatial approach, they should know about this, both for ensuring the effectiveness and for securing the legitimacy of spatial solutions in flood governance (Doorn 2016). Thus, the question of "who pays" for spatial measures is a very relevant one, which deserves attention in the scientific literature on flood risk management. This book further explores this angle.

1.1.3 Two Extreme Cases: The Netherlands and the United States

This book analyzes the shift to spatial measures in flood governance in two national cases: the Netherlands and the United States (US). In both countries, a shift to spatial measures materialized, but this shift took place in a very different time period and had different (distributive) results.

The Netherlands has a long tradition of flood prevention. This delta country partly came into existence by reclaiming land from the sea. Over the years, the Dutch national government built impressive levee systems to continue to ensure safe living conditions within its territory. Since the mid-1990s, spatial measures have been incorporated into Dutch flood governance. In the US, a transition from safety to spatial measures already occurred in the 1960s. Practically embodied in a federal National Flood Insurance Program (NFIP), this spatial approach institutionalized over time and still functions as the primary framework through which floods are governed.

In this sense, the Netherlands and the US can best be seen as two "extreme" cases of the shift to spatial measures that currently progresses in many flood-prone countries. According to Yin (2009: 52), extreme cases may help to "reveal insights about normal processes" that are harder to detect in "typical" cases (see also Flyvbjerg 2006; Seawright and Gerring 2008). In this book, the extreme case setup was selected to shed light on the varying nature of distributive choices in flood governance.

1.1.4 Research Aim

By analyzing the shift to spatial measures in two extreme cases, the aim is to generate insight in the variety of policy approaches and distributive implications that can underlie a spatial approach to flood governance. In addition, this book aims to improve our knowledge of the process through which the allocation of costs and responsibilities for spatial measures are determined. The provisional literature on this topic suggests that multiple cost and responsibility allocations are possible under a spatial approach. However, how do policymakers decide which allocation scheme to use? Insight into these processes can help policymakers in countries that are currently seeking to integrate spatial measures in standard governance approaches to floods to design (more) effective and legitimate spatial policies in flood governance.

1.2 Theoretical Framework

That the distributive implications of spatial measures not received much attention in the academic literature thus far, can probably partly be explained from the huge diversity that underlies the shift to spatial measures in different countries. While many countries have integrated spatial measures in their portfolio of flood risk management strategies, they have done so to a different degree and with a huge variety in measures and the organization of these measures in terms of governance (Hegger et al. 2016, 2017). Spatial measures are implemented in existing national and regional (flood) governance contexts, which are embedded in historical traditions of dealing with floods. This variety and contextual embeddedness complicate the generation of general conclusions about the distributive implications of a shift to spatial measures in flood governance; cost and risk distributions will differ from country to country and from region to region, depending on specific management choices in particular (historical) contexts.

Another explanation of the meager attention to the distributive implications of spatial measures may lie in the fact that up to now, research on spatial measures in flood governance has mostly found an uptake in the field of environmental economics. Within this field, environmental problems are studied from a problem perspective focused on efficiency and risk distributions are also analyzed from this perspective. In this section, the environmental economics perspective is briefly introduced, along with other possible perspectives to analyze distributive implications in flood governance. The section then zooms in on the political perspective and a particular challenge that arises from this perspective for distributive decision-making on floods, which will further guide the analysis in this book.

1.2.1 Different Research Perspectives on Flood Risk Distributions

An Economic Perspective: Efficient Distributions

For the academic field of environmental economics, the key question is what risk distributions are efficient, that is, when environmental risks are distributed in such a way that the total sum of costs for protection and damage is minimized. Safety standards for levees are, for example, often based on environmental economic analyses. Safety standards are determined by comparing the investments needed to build or strengthen and upkeep levees with the damage mitigated by these investments. The point at which investments in structural flood defenses no longer outweigh the expected damage averted by these investments, flood protection is no longer "rational". Environmental economics does not only look at the financial side; other aspects are factored in as well, such as the value of life, the natural and cultural value of land, and the impacts of structural flood defense measures on nature and economic sectors such as shipping and international trade.

Two factors that have received particular attention in environmental economic analyses are climate change and urbanization. Climate change changes flood probabilities. Sea-level rise, increased snowmelt, extreme rainfall, and high wind speeds put pressure on existing riverine and coastal flood defenses, increasing the risk of these defense systems to be overtopped or breached, with flooding as the result. At the same time, the on-going urbanization in delta regions increases the value exposed to floods, resulting in a huge disaster when a flood strikes. The enormous havoc caused by Hurricane Katrina in Gulf Coast countries in 2005 provides a vivid example.

Climate change and urbanization alter the trade-off between investments in structural flood protection and the damage averted. The incorporation of these factors in environmental economic analyses has led to the recognition of a cycle in which societies invest in stronger flood defenses to protect growing settlements in flood-prone areas, drawing only more developments to these areas, requiring even stronger flood defenses. From this recognition, environmental economists have argued that spatial measures can provide a cost-efficient alternative. Spatial measures minimize the value exposed to floods, and can, therefore, eliminate the need for expensive flood defenses in the first place (Wolsink 2006; Hartmann 2009).

Environmental economists usually do not argue for a complete turnover to spatial measures in flood governance. Rather, they call for the consideration of spatial measures as part of a more "risk-based" approach to floods, in which the costs and benefits of structural flood protection are weighed off against the costs and benefits of alternative solutions like spatial measures (Sayers et al. 2002; Botzen and Van den Bergh 2008; De Vries and Wolsink 2009; Liao 2012). In a risk-based approach, spatial measures can co-exist with structural flood defenses, the weight placed on each solution being dependent on the total sum of costs and benefits generated by their balance.

An environmental economics perspective helps to identify and examine the costs and benefits involved with different environmental risk management strategies, including those devised for floods. These costs and benefits are evaluated, however, only to the extent that they contribute to efficiency; certain distributions are more "efficient" than others. Environmental economists do not make an evaluation of the impacts of these distributions for different social groups. These questions fall into the scope of environmental ethics.

An Ethics Perspective: Fair Distributions

In environmental ethics, the key question is what environmental risk distributions are just or fair. Theories of distributive justice are used to come to a well-considered normative judgment of different allocations of costs and responsibilities in environmental governance. Within distributive justice theory, two normative accounts can be distinguished: a desert-based and a consequentialist account (Low and Gleeson 1998; Shue 1999; Paavola and Adger 2002; Dellink et al. 2009).

The "desert-based" account of distributive justice departs from Kant's principle of just intentions, which holds that people are rational and can, therefore, be held responsible for their actions. Following this principle, desert-based accounts claim that distributions are just when burdens and benefits are allocated to those at the cause. An example of the application of this account is the "polluter pays" principle, which holds that the "wrongdoer" should pay for the environmental (side) effects caused by its actions.

Consequentialist accounts do not focus on the (rational) actions that produce burdens and benefits but look at the outcomes of cost and benefit distributions in society. A further distinction can be made between egalitarian and utilitarian consequentialist accounts. Egalitarian consequentialism aims for distributions that in effect reduce existing inequalities between social groups of regions. Distributions for example follow Rawls' "maximin" principle which involves maximizing the well-being of those who are worst-off. Utilitarian consequentialism is based on Mill's and Bentham's "greatest happiness principle", and distributions aim to maximize the positive effects and minimize the negative effects on society as a whole.

Doorn (2016) emphasizes the importance of considering fairness in distributions of responsibility for flood governance, both from an ethical and an instrumental point of view. Ethically, it is important that these distributions are just and considered legitimate by society. From an instrumental point of view, fair distributions that are considered legitimate lead people to take up their responsibility. This becomes more important when responsibilities are decentralized and deregulated, as can be the case under a spatial flood governance approach. An environmental ethics perspective can clarify the normative underpinnings of a spatial approach to floods and, as such, help to justify, legitimize and implement spatial policies in flood governance.

However, the normative evaluations that environmental ethics support, can lead to different outcomes. That is, people can arrive at different conclusions about which distributions of costs and responsibilities are "right" or "just" in flood governance. While it is important that the normative underpinnings of different policy approaches to floods are clear and their distributive effects are well understood, the final choice regarding what distribution to follow is inherently political.

A Political Perspective: Legitimate Distributions

From a political perspective, what is principally at stake is not a normative evaluation of environmental distributions but the processes through which these distributions are produced. As Lasswell (1936) described politics as the process of determining "who gets what, when, and how", the allocation of burdens and benefits over members of society is seen as the core of political decision-making. It is in these processes that legitimacy can be gained (or lost) for certain distributive choices (Rawls 1999).

In democratic political systems, legitimacy can be gained by following standard democratic principles, such as representation and deliberation. For example, a central democratic premise is that "political" decisions should be made by representative decision-making bodies. Representation warrants an equal distribution of power; it ensures the viewpoints and interests of different groups are taken into account in public decision-making (Dahl 1998, 2006). Just as important, democratic principles call for an open dialogue in which policy preferences can be publicly contested and discussed (Fennema 1982). Through this dialogue, policy alternatives can be evaluated from different (normative) perspectives, which inspires collective learning and contributes to the legitimacy of policy choices because underlying trade-offs have become clear and have been justified in open and transparent deliberations (Estlund 2008).

In his elementary treatise on representative government, John Stuart Mill (1861) famously applied these principles as "checks" in representative democratic systems. In Mill's model, the government is run by experts, capable of coordinating complex state affairs. This government is checked by an evenly competent, representative parliament capable of following and judging the actions of governors. Democratic elections warrant representation in parliament and, therefore, Mill was a strong advocate of extending the national suffrage to all citizens. According to Mill, parliament is "a place where every interest and shade of opinion in the country can have its cause even passionately pleaded, in the face of the government and of all other interests and opinions, can compel them to listen, and either comply, or state clearly why they do not" (idem.: 226). This representative body, in turn, is controlled by "public opinion", institutionalized through democratic elections.

Deliberation takes place in the process of parliament controlling the government. According to Mill (1861: 226), the parliament's main function is "to throw the light of publicity on its [the government's, EB] acts: to compel a full exposition and justification of all of them which any one considers questionable; to censure them if found condemnable, and if the men who compose the government abuse their trust, or fulfill it in a manner which conflicts with the deliberate sense of the nation, to expel them from office, and either expressly or virtually appoint their successors". Thus, in the process of controlling the government, elected representatives call on governors to explain their rulings, evaluate and judge these explanations in dialogue with other representatives and governors, reserving the right to remove them from office when their actions are no longer in line with public viewpoints or trust.

When thinking about distributing costs and responsibilities in flood governance, a political perspective emphasizes the importance of looking at the decision-making processes through which distributive decisions are produced. Democratic representation and deliberation are highlighted as key principles to generate political legitimacy for these decisions. In modern representative democracies, these principles have been organized along the lines of Mill's representative model. However, this organization has undergone severe criticism since the 1970s. An important point of criticism focuses on the involvement of experts in democratic decision-making.

1.2.2 The Expert-Democracy Tension in the Political Perspective

One line of critique deals with the position of experts in representative decision-making bodies. Mill was in favor of an educated parliament, proficient enough to watch over, judge, and control the actions of government professionals. Today, highly-educated groups are still overrepresented in national parliaments. For some, this overrepresentation is principally undemocratic as it undermines public representation (Bovens and Wille 2011). Mill's model has been defended against this critique by re-emphasizing the equal right to vote as an equal *public check* on those in parliament (Ankersmit 1999; Manin 2007; Rosanvallon 2008). This discussion is part of a fundamental debate in political theory, which will not play a big part in this book. Representative institutions are accepted as democratic. Rather, the focus will be on the role of experts in the policymaking process leading up to political decision-making, which is part of a second line of criticism on the role of experts in modern democratic systems.

This second line of criticism started with developments in the field of Science Studies, and the subfield of Science and Technology Studies (STS) in particular, which demonstrated that knowledge and values are strongly interconnected; rational knowledge is not value-free but underpinned by normative assumptions and prepositions of how the world operates and how it should operate (Jasanoff 1990; Nowotny et al. 2001). However, the knowledge of experts is often accepted as "true" in democratic decision-making (Estlund 2008). Even when experts are only involved in "executive" or "advisory" branches of government, their influence can reach into the political domain, reducing room for the formulation and inclusion of counter-perspectives and policy alternatives, and undermining the open and inclusive character of democratic deliberations and decision-making (Habermas 1996).

This expert-authority is not very problematic for coordination problems, which for example play up in traffic management. Here, the challenge is to find the most efficient solution while values and interests play no particular part (Engelen 2008). Problems with expert-influence do arise in the domain of distributive problems, where a trade-off has to be made between different interests and perspectives. These problems, as Habermas (1996: 429) famously stated, "cannot be reduced to the inefficiency of administrative steering". They should be resolved through democratic decision-making procedures, which ensure that different perspectives, values, and interests have been explicated and traded off (Rawls 1999). However, distributive problems are not always clearly delineated; especially when they are surrounded by scientific uncertainty, distributive questions can be hard to identify while it is especially in these areas that the reliance of expert-knowledge is high (Rittel and Webber 1973; Hisschemöller and Hoppe 2001; Dijstelbloem 2007).

Over the years, political scientists have identified strategic uses of expert-knowledge in democratic decision-making. They have demonstrated that claims to expertise can be "organized in" the democratic decision-making process to support certain views or interests, whereas other claims (and other views and interests) can be "organized out" (see, for example, Weiss 1979; Jasanoff 1990; Hajer 1995; Turner 2001; Maasen and Weingart 2005; De Bont and Vanpaemel 2012; Jennings and Hall 2012; Wesselink et al. 2012; Elgert 2013; Bock 2014; Dunlop 2014; Javeline and Shufelt 2014; Lundin and Öberg 2014; Metze 2014; Rietig 2014; Waylen and Young 2014; Rimkutė and Haverland 2015).

For political scholars, the revelation *that* expert-influence often trespasses its boundaries and invades the "political" domain of decision-making has been reason to theorize about new forms of "participatory" or "deliberative" democratic governance (Barber 1984; Hoppe 1999; Warren 2006; Bader 2014) or new forms of democratic accountability (Majone 1999; Rosanvallon 2008; Bäckstrand et al. 2010; Borowiak 2011) to better counterweigh and check the influence of experts. Strikingly, however, the literature has paid little attention to understanding *how* experts actually influence distributive decision-making in representative democracies: *how* experts influence the process of finding fair allocations of burdens and benefits in democratic decision-making, *how* their knowledge impacts the formulation and evaluation of alternative perspectives in this process, *how* expert influence relates to that of "political" actors in this regard and *how* through this relationship the distributive aspects of governance aproaches are shaped.

1.2.3 Expert-Democracy Tensions in Risk Governance

Tensions between democratic and expert-form of governance are amplified in risk management. In their seminal writings, Van Gunsteren (1976) and Scott (1998) highlighted the increased dependency on expert-knowledge in modern societies' "quest for control", while simultaneously showing the fallibility of expert-knowledge to deal with distributive challenges. Risk scholars argue that this quest for control reaches new levels in risk governance.

Increased prospects for predicting and controlling risks have increased the public demand to be protected against risks (Reddy 1996). This trend has been signaled by scholars working on the "risk society", a term used to describe the preoccupation of modern societies with risk control (Beck 1992; Giddens 1999). In risk societies, political legitimacy is derived from the ability to control risks and therefore, policymakers strongly rely on experts to devise risk management strategies (Beck 1995). Risk scholars argue that this expert-knowledge can miss its target. Risks are standardly defined as “probability x consequences” (Renn 1992) and expert-solutions typically aim to minimalize the outcome of this equation. In reality, however, risks are more complex. Their management poses a collective action problem because the burdens of investing in risk protection and the benefits received from this protection are generally not equally spread (Reddy 1996). Policymakers have to decide on the acceptability of (the costs that are needed to prevent) risks, as well as on the distribution of the costs and benefits of risk protection in society (Kane and Shorgen 2000). Rational risk calculations do not help policymakers with these choices. To the contrary, risk scholars have argued that by reducing complexity surrounding risks, risk calculations “depoliticize” risk governance (Beck 1995).

Insights about the partial and political character of expert-knowledge in risk governance have led to a decline of trust in risk expertise. Rosanvallon (2008: 9) pinpoints this problem as follows: “The risk society is by its very nature wary of the future, yet its citizens are still obliged to place their trust in scientists because they cannot weigh the relevant issues without the aid of specialists. This role of scientists is as problematic as it is indispensable, and this is a source of resentment. Citizens have no alternative but to oblige scientists to explain their thinking and justify their actions.” Risk scholars have, therefore, also stressed the need for more participatory and deliberative decision-making procedures (Hoppe and Petersen 1993; Adam et al. 2000; Kahan 2006; Renn and Schweizer 2009; De Marchi 2015) and new forms of democratic accountability (Rosanvallon 2008; Cutler 2010) in risk governance.

1.2.4 Research Question and Conceptual Clarifications

The shift from safety to spatial measures can substantially alter traditional cost and responsibility distributions in flood governance. But who decides on new responsibility structures under a spatial approach? From a political perspective, it is important that these decisions are made through standard democratic procedures that ensure the legitimacy and effective implementation of distributive choices. In these procedures, experts can be involved in “executive” and “administrative” domains of flood governance where public policies are implemented, but their influence should not reach into the “distributive” domain of decision-making where choices are made about the division of costs and responsibilities for dealing with floods. Such distributive decisions are explicitly reserved for democratic decision-making bodies.

However, the recognition of the political nature and natural authority of expert-arguments in public governance, and in risk governance in particular, has put this standard democratic model to the test; time and time again, political analysts have revealed that experts can—willfully or unintentionally—influence distributive decision-making. Political scientists have called for the organization of new, more participatory or deliberative forms of decision-making and accountability to sustain political legitimacy in risk governance.

At the same time, the reliance on expert-knowledge in flood governance is high and this is for good reasons. Floods pose an imminent threat to societies around the world. Lives and capital are directly at stake. Flood governance requires a good balance between expert and democratic forms of governance, to ensure both the effectiveness and the legitimacy of flood risk management. A critical question becomes: To what extent can we trust and under what conditions do we allow experts to make decisions for us, especially in the distributive domain of flood governance?

To organize a good balance between expert and democratic elements in flood governance, a good understanding of the relationship between experts and political actors in, and the connected effects on, distributive decision-making on floods is needed. Are the normative assumptions behind expert-arguments sufficiently recognized by political actors in democratic decision-making processes on floods, and are they sufficiently tested and challenged by alternative views in democratic deliberations?

This question will be analyzed in the two cases of the shift to spatial measures in Dutch and US flood governance. The research question is: **What experts were involved in the shift from safety to spatial measures in Dutch and US flood governance, and how have these experts influenced the distributive decision-making process underlying this shift?**

This research question requires two points of conceptual clarification. First, it is good to specify the use of "decision-making", especially in relation to the related concept of "policymaking". In traditional democratic models, political "decisions" are produced by democratic decision-making bodies that warrant an equal consideration of different views and interests in the decision-making process (Bovens et al. 2012). These decisions are translated into "policies" by independent state administrators in the executive branches of government. In reality, the boundaries of this politics-administration dichotomy are blurred (Zanetti and Adams 2000; Engelen 2008). In contemporary democratic accounts, decision-making is mostly conceptualized as taking place in all stages of the "policymaking process", from agenda-setting to policy implementation (Hupe and Hill 2006). Drawing on these insights, this book uses "policymaking" to refer to the whole process through which policy problems are identified, put on the agenda, and addressed through the formulation of public policies. When "decision-making" is used, this will refer in a more classical sense to the narrower process through which policymakers reach a decision on the distribution of burdens and benefits, as well as the allocation of costs and responsibilities, in flood governance.

Second, the use of the terms "expert" and "expert-knowledge" should be explained. Existing studies on expert influence often use predefined categories to indicate which actors count as "expert actors" or which knowledge counts as "expert knowledge" in their analyses (e.g., Maasen and Weingart (2005) look at scientific advice committees). At the same time, political analyses on expert-influence have demonstrated that the boundaries between "politics" and "expertise" are very thin and inherently contested (Metze 2014). Therefore, this book builds on the more empirically-grounded understanding of expertise brought forward by authors like Jasanoff (2004), Hajer (1995), and De Swaan (2004), which accepts that claims to expertise can be organized by actor-groups to highlight the relevance of their expertise in public policymaking and that these claims have to be recognized by other groups in order to be of influence in the policymaking process.

Last, the term "spatial measures" requires some further clarification. Especially because safety measures also have an important "spatial" component, not only in the sense that they are physical structures that are strategically located to defend certain (highly-valued) areas against flooding, but also because structural flood defenses like levees and floodwalls often have massive implications for the landscape in which they are placed. In this book, however, the term spatial measures is used to refer to a (new) type of flood governance measure that distinguishes itself from safety measures by its purpose. Whereas safety measures aim to keep the water out of the physical landscape, spatial measures intent to adjust the spatial planning structure of the physical landscape to accommodate floodwater without causing damage.

1.3 Conceptual and Analytical Framework

This section outlines the conceptual and analytical framework used to analyze the research question formulated above. Conceptually, the shift to spatial measures in flood governance is grasped in terms of a process of gradual institutional change, for which the Policy Arrangements Framework (PAF) provides the necessary conceptual tools. The influence of experts in this process is analyzed through the analytical notion of framing. Based on this framework, research sub-questions are formulated.

1.3.1 Institutional Change

In this book, the shift to spatial measures is understood as a process of institutional change. At a very basic level, institutions refer to patterns that structure social life. German sociologist Max Weber (1978: 23–31) referred to them as "social orders". By setting out shared norms for and expectations of social behavior, institutions enable social interaction. Institutions structure interaction, but they are also reproduced through these interactions (Giddens 1984). They, therefore, have a structuring power that tends to remain stable over time, especially when they become part of formal organizational structures like those in government (DiMaggio and Powell 1983; March and Olson 1989).

From this basic understanding of institutions, different strands of institutional theory have been developed in different research fields. Historical institutionalism, for example, focuses on explaining the durability and change of large-scale socio-political systems, such as communism or certain forms of democracy (Skocpol 1979; Lijphart 1999). Institutional economics or rational-choice institutionalism analyses patterns in economic behavior (Hindmoor 2010). Sociological institutionalism aims to understand how institutions are shaped and reshaped in their particular (political, cultural, economic, etc.) context (Lowndes 2010).

This book builds on the sociological strand of institutionalism. It focuses on institutions in flood governance, which are the basic structures through which a society thinks about and deals with floods. They forward "logical" rules for cost and responsibility allocations in flood governance. These institutions are embedded in a wider historical, political, cultural, economic, and physical context. They emerge from this context, are reproduced in this context, and change in this context.

Institutional change has long been explained by changes in the external context that challenge the "logic of appropriateness" of prevailing institutions (March and Olsen 1989: 22). For example, shifts in the political world order have been used to explain the rise of social revolutions (Skocpol 1979) and in environmental governance, shock events like a flood have been linked to occurrences of institutional change (Birkland 2007). Institutional change has, therefore, often been conceptualized as a "big shift" or "turn", changing not only existing policy approaches but also the deeper-lying assumptions, norms and values in a policy field (Van Tatenhove et al. 2000; North 1990).

Increasingly, the role of agency in bringing about and shaping institutional change has been highlighted as well (Clemens and Cook 1999; Arts and Van Tatenhove 2004). Studies have shown that external changes and shock events can be strategically used by actors with an interest in changing their institutional environment, and reversely, that actors use contextual factors to upkeep the institutional status-quo (Hajer 1995; Birkland 2007). Most theories on institutional change now incorporate contextual and agency factors in their explanations, aiming to better understand how both factors relate to each other in processes of institutional change (Hodgson 2007). Agency is usually located in moments of big institutional transformation. In his "three streams model", Kingdon (1995), for example, poses that the combination of contextual changes that undermine existing governance approaches, and the existence of policy actors that develop new solutions and organize the political will to implement those solutions, opens a "window of opportunity" for institutional change.

The understanding of institutional change as a complete turnover from one institution to another has been questioned. Some scholars have called for a more nuanced perspective on institutional change, in which institutions do not necessarily completely transform—although this is still possible—but rather gradually evolve in response to incremental, contextual developments and processes of political agency in daily political practices (Mahoney and Thelen 2010; Gray et al. 2015). In this perspective, institutional change is conceptualized as a process of gradual "institutional adaptations" in which context and agency factors are constantly at play, pushing for and against institutional change (Hall 1993; Hall and Taylor 1996).

In this book, the shift from safety to spatial measures in flood governance will be analyzed as a process of gradual institutional adaptation. This means that rather than conceptualizing institutional change as a fixed moment in time where contextual factors and agency collide into the collapse of existing and build-up of new institutions, institutional change is understood as a lengthy process in which institutions gradually evolve. This evolutionary perspective on institutional change implies that agency is asserted constantly, in daily public policymaking, where actors do not necessarily try to change the whole institutional framework but influence its different subparts, leading to gradual adaptations of the institutional framework over time.

1.3.2 The Policy Arrangements Framework

The evolutionary perspective on institutional change has been elaborated on in the Policy Arrangements Framework (PAF). In the PAF, long-term institutional structures in a policy field are linked to daily policymaking processes in that field (Van Tatenhove et al. 2000; Arts and Van Tatenhove 2004; Arts et al. 2006). Policy arrangements are defined as "temporary stabilizations in the content and organization of a policy domain" (Arts et al. 2000: 54); they are the institutionalized values, norms, perceptions, and practices in a policy field. By emphasizing the temporary character of policy arrangements in this definition, change is seen as an inherent part of institutional development in public policymaking. The PAF sets out a framework for analyzing these institutional developments.

A policy arrangement has two main dimensions: a substance and an organizational dimension. The substance of a policy arrangement is formed by the policy discourse, defined by Hajer (2005: 300) as the collection of "ideas, concepts, and categories through which meaning is given to social or physical phenomena and which is produced and reproduced through an identifiable set of practices". In the PAF, these practices are part of the organizational dimension, which is subdivided into an actor, a resources, and a rules dimension. The actor dimension captures the composition of actors and actor-groups involved in a policy field. The resources dimension arranges the allocation of power and resources between these actors. The rules dimension specifies the rules and procedures based on which policy actors interact in the policy field. Figure 1.1 depicts the policy arrangements framework schematically.

The distribution of costs and responsibilities in a policy field is shaped by the institutional structure of the policy arrangement. The policy discourse sets out the basic distributive principles that underlie public policymaking in the policy field. These principles structure who is involved in the governance of the problem at the actor dimension, how costs and responsibilities are divided between them at the resources dimension, and which rules and procedures apply to these distributions at the rules dimension. Thus, policy arrangements set out the political structure in a policy field, in a Lasswellian sense, that determines who gets what, when, and how (Arts and Van Tatenhove 2004).

In principle, the institutional structure provided by a policy arrangement is stable. However, the framework explicitly accounts for change. In a policy arrangement, change can originate in each dimension; however because all dimensions are interrelated, change at one dimension can prompt shifts in the other dimensions as well. At the same time, the structural features of the policy arrangement determine the opportunities for and directions of change in each dimension (Arts and Van Tatenhove 2004: 5). In the PAF, institutional change is thus conceptualized as a gradual process where the different dimensions not necessarily change all at once but gradually and in response to each other, which may or may not cause the whole framework to “tilt”. The extent and speed of these developments are determined by the interplay between internal and external pressures at each dimension.

Expert influence has been linked to the actor dimension. Experts are one of the actors involved in public policymaking (Wiering and Immink 2006: 425). Depending on their specific role in the policy field, experts influence the policy discourse that emerges in a policy field, which further transpires in the organizational dimensions of the PAF: the actor, the resources, and the rules dimension. Through these organizational dimensions, the distribution of costs and responsibilities is shaped.

Being part of the policy arrangement, the influence of experts is also subject to institutional development and change. On the one hand, there is a strong, internal push for stability. Expert influence can be upheld by prevalent policy discourses that bring forward certain types of expertise as “logical” or “common sense”, by vested actors with an interest in the status quo of science-policy relations and the resources to maintain this status, or by the “rules of the game” which grant certain types of expertise a preferential status in policy procedures. On the other hand, expert-influence can be challenged and changed by developments in the context policy field that contest the standing policy discourse and may lead actors to question the science on which it rests. But this change can also originate within the policy field, when actors question the efficacy of the existing policy discourse and the rules that stem from it, or the institutionalized science-policy relations that underlie it. In a policy arrangements perspective, expert-influence is constantly (re)defined in relation to the other dimensions of the policy arrangement.

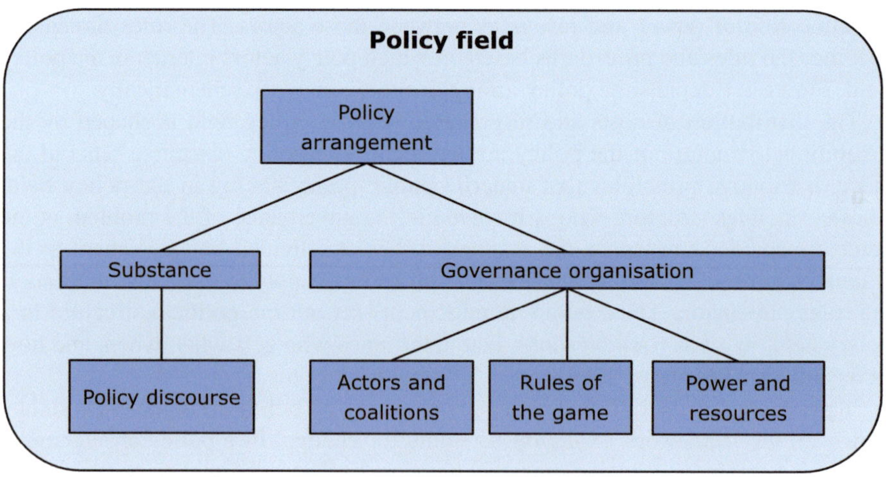

Fig. 1.1 The policy arrangements framework (adapted from Arts et al. 2006: 99)

1.3.3 Framing Theory

The conception of the shift to spatial measures in flood governance as a process of gradual institutional adaptation helps to locate expert-influence in, and the effects of expert-influence on, the policymaking process underlying this shift. However, it does not explain *how* experts may influence institutional adaptation, and the distributive rules that underlie it. For this, framing theory provides a useful basis.

The concept of a frame finds its origins in individual psychology, where frames are seen as representations through which individuals make sense of the world around them. Generally, a distinction is made between cognitive and communicative framing theories (Van Hulst and Yanow 2016). In cognitive approaches, frames are located in the individual's mind. They are built up from past experiences and change when the individual incorporates new understandings gained from new experiences. In communicative variants, frames are "shared" collective entities. They are the "primary frameworks" that people draw on to make sense of a situation and their role in it (Goffman 1974).

The communicative variant is used in public policy analyses. In these analyses, public policymaking is conceived of as a discursive struggle between different actor-groups who try to generate support for their policy ideals by presenting them in a logical and coherent story about a policy problem that "sounds right" or "rings true" (Rein and Schön 1993). These stories are referred to as policy frames.

By presenting a "simplified" picture of a multifaceted public policy reality, policy frames transform complexity in public governance into "structured and meaningful policy problems" (Rein and Schön 1993; Verloo 2005: 20). Policy frames, therefore, have an important enabling function in public policymaking; they help actors come to grips with complex policy situations based on which policy strategies can be devised (Maussen 2009). However, the scope of policy frames is necessarily also restrictive. As one problem perception starts to dominate the policymaking process, alternative interpretations can be excluded.

The characteristics that make certain frames "stronger" than others, and resonate better in public policymaking, have been studied (Gamson and Modigliani 1989; Benford and Snow 2000). Benford and Snow (2000) identified a number of features that contribute to frame resonance. The inclusiveness of policy frames (i.e. the extent to which policy frames are open to different interpretations and can, therefore, encompass different problem perspectives) is one of these features. Salience is another factor, which refers to the extent to which a policy frame fits with people's culture, beliefs and experiences (on cultural resonance, see also Gusfield 1981). They also highlight credibility as an important factor contributing to the strength of a policy frame. Credibility depends on frame consistency (is the same storyline told over time?), practical reliability (does the storyline fit with developments in practice?) and the credibility of frame articulators. The credibility of frame articulators greatly depends on the (trustworthiness of) experts the committed to the policy frame.

For this reason, policy frames have been recognized as powerful political instruments (Gamson and Modigliani 1989; Rein and Schön 1993; Benford and Snow 2000;

Hajer 2005). When actors communicate their views about a policy problem—an activity referred to as framing (Benford and Snow 2000: 614)—they shape other people's views on the nature of the problem and the appropriateness of certain policy solutions. In political accounts, therefore, framing is seen as a third form of power as identified by Lukes (2005). The first form of power is exercised through conflict. It for example becomes visible in the endeavor between different political parties and interest groups to get their preferences translated in public policies (see also Dahl 1961). The second form of power is based on exclusion, which for example plays up when certain interest groups are excluded from agenda-setting and lobbying arenas (see also Bachrach and Baratz 1962). In comparison to the first two forms of power, the third form of power distinguished by Lukes is less easy to observe. It works through the generation of consent with the preferences and ideas of certain (often dominant) groups in society, and, therefore, remains invisible even to those subordinate to this form of influence.

In political accounts, much attention has been paid to the strategies used by actor coalitions (or frame supporters) in the production and development of policy frames. Knowledge forms an important part of these strategies. Drawing on Sabatier and Jenkins-Smith's (1988: 158) often-recited insight that public policies "are based upon (often implicit) causal theories of how the world operates, and […] much of the policy debate can be understood as disputes over the validity of those causal theories", Fischer (1995: 111) has argued that all policy frames are "layered" in the sense that value orientations always underlie rational explanations of a policy problem. Influential theories, such as Sabatier and Jenkins-Smith's (1988) Advocacy Coalition Framework and Hajer's (2005) Discourse Coalition Framework, pose that actor coalitions are closely entwined with "epistemic communities", which Haas (1992: 3) defined as "network[s] of professionals with recognized expertise and competence in a particular policy domain and an authoritative claim to policy relevant knowledge within that domain or issue-area" (for a practical example, see Meijerink 2005). Political framing accounts have typically deconstructed the discursive tactics used by different actor-groups in their struggle for discursive hegemony by unraveling the intimate relationship between experts and political actors in this process (Gusfield 1981; Hajer 1995; Jasanoff 2004; Metze 2014).

Traditional political framing accounts have been subject to criticism and revision. In their influential critiques, Entman (1993) and Steinberg (1998) argued that framing scholars tend to see public policymaking too much as a disconnected process in which stable actor groups with unchanging views and interests compete for discursive hegemony. They posed that framing always takes place in a larger social, political and cultural context where existing policy discourses and political ideologies not only hugely impact the policy frames that are produced by actor groups but also the extent to which these policy frames are accepted by other policy actors. These views have inspired the development of a new, "interactional" account of framing. In this account, public policymaking is no longer seen as a discursive battle between the policy frames put forward by different actor groups but as a joint process of learning and meaning-giving, directly in response to new policy situations, in which the policy frames of different actors "interact" (Dewulf et al. 2004, 2007, 2009). In this process, agency and collective sense-making are closely entwined (Van Hulst and Yanow 2016). The policy frames that emerge from this process are "co-constructions of meaning", produced by the interaction between political and rational (including expert) arguments (Dewulf and Bouwen 2012: 169).

Framing theory provides an explanation of how actors influence public policy-making. That is, they use narratives and storylines to generate consent with their views on the nature of a policy problem and the best policy solution to deal with this problem. Experts play an important role in this framing process. They offer a knowledge base that supports the normative views of actor coalitions.

Recent framing accounts highlight the contextual nature of the framing process. Cultural, historical, social and political factors do not only determine which policy frames are developed but also how they are received. In addition, when the frames of different actor coalitions interact in concrete situations of public policy making, strategic agency and collective sense-making collide, producing new, "shared" policy frames in the policymaking process.

1.3.4 Analytical Framework and Sub-questions

Following the research question set out above, the specific research objective of this book is to contribute to a better understanding of the relationship between experts and political actors in, and the effects of this relationship on, distributive decision-making underlying the shift from safety to spatial measures in Dutch and US flood governance. The theories and perspectives discussed in this section offer the conceptual and analytical tools to structure the analysis.

In this book, the shift from safety to spatial measures in Dutch and US flood governance is understood as a process of gradual institutional adaptation. In this perspective, flood governance institutions do not change all at once but gradually evolve in response to short-term internal (agency) and external (context) factors. The PAF conceptualizes these institutions as policy arrangements with different dimensions—a policy discourse, an actor, a resources and a rules dimension—each dimension being subject to internal and contextual pressures for stability and change. Institutional adaptation can, therefore, take different forms and occur at different speeds in different (national) settings, with different implications for the distribution of costs and responsibilities in each setting.

In the literature on spatial measures in flood governance, there is a discussion on the extent to which the introduction of spatial measures actually encompasses institutional change. Some studies find that spatial measures change the institutional underpinnings of flood governance (Pahl-Wostl 2007; Jong and Van den Brink 2013) whereas other studies conclude that old policy institutions largely remain under a spatial approach to floods (Wiering and Arts 2006). By grasping the shift to spatial measures as an evolutionary process of gradual institutional adaptation, the variety of these conclusions can be accounted for (see also Wiering et al. 2017). This book approaches this shift as a gradual, incremental process, which, in some countries, gathers more pace than in others and evolves differently in each setting.

From an institutional perspective, actors can influence their institutional context. An evolutionary perspective on institutional change implies that power and influence are constantly at play. Framing theory helps to understand how this influence works. It holds that as policy actors interact, they co-produce policy frames that influence the policy discourse that structures democratic decision-making in a policy field. However, opportunities for actors' influence are structured by the same institutional context in which they operate. Figure 1.2 depicts the relationship between the PAF and framing.

Combined with the PAF, framing can be used to analyze how in the actor dimension of the policy arrangement in Dutch and US flood governance the arguments of experts influence the policy discourse on spatial measures, what (new) distributions of costs and responsibilities for flood protection were produced in this process, and how the rules that underlie these distributions change as a result. This influence should be understood against the background of the existing institutional and external context in a policy field. Such an analysis calls for a long-term, empirically grounded research approach that is able to grasp the structural features of a policy arrangement in relation to the daily internal and contextual developments in a policy field that shape institutional structures. The analysis in this book, therefore, adopts a long time-horizon, studying the implementation of spatial measures in the Netherlands and the US from the first startup to its recent developments, looking at how experts, in interaction with political actors, influenced this institutional path and its distributive underpinnings.

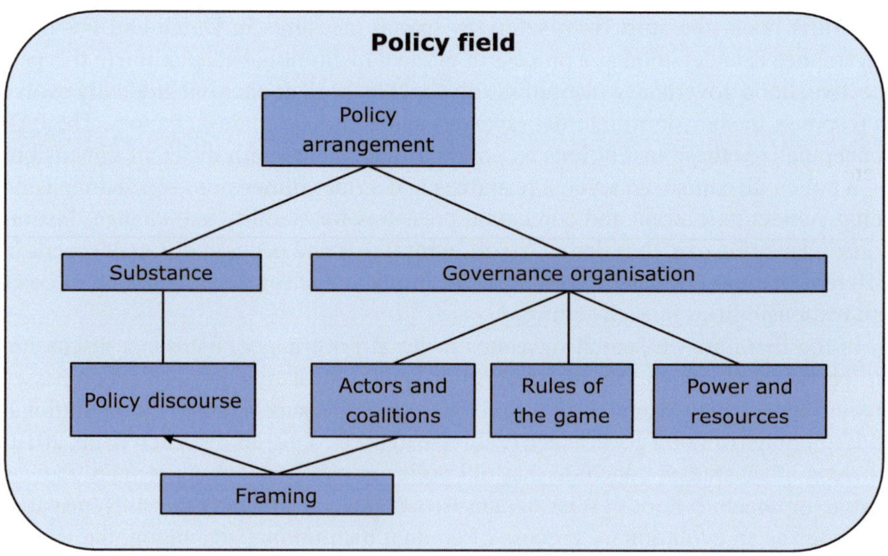

Fig. 1.2 Framing and the policy arrangements framework (adapted from Arts et al. 2006: 99)

The following sub-questions have been specified:

1. How has the shift from safety to spatial measures in flood governance been framed in the Netherlands and the US?
2. To what extent and how has the policy arrangement underlying Dutch and US flood governance, including its distributive aspects, changed as a result of these framings?
3. Which experts were involved in the policymaking process on spatial measures in Dutch and US flood governance and how have they influenced the framing and institutional adaptation processes in both countries?
4. What did this influence of experts imply for the recognition of the distributive implications of spatial measures in the democratic decision-making process?

1.4 Data Collection and Methods

Policy change—institutional change in particular—is a lengthy process (Hajer and Laws 2008: 264, Van Hulst and Yanow 2016). To fully grasp the processes underlying and the changes brought forward by the shift from safety to spatial measures in Dutch and US flood governance, the cases analyzed in this book cover a long period. They start out with an analysis of the institutionalization of the safety approach in the late 19th century and end with an analysis of recent, spatial policy developments. The case studies outline general developments in Dutch and US flood governance over this period, and then they zoom in on several key moments in the formation or change of policy approaches in both countries. For these moments, the policymaking process was reconstructed.

In these policymaking reconstructions, expert-influence was traced back by looking at how the arguments of experts, in relation to the arguments of other policy and political actors, shaped the policy discourse on spatial measures in Dutch and US flood governance, and how through this, the distributive aspects of spatial policies were formed. This analysis is based on the method of discourse analysis, which studies how actors give meaning to social and physical phenomena to explain (the outcomes of) processes of social interaction (Parsons 2012) and which is often used in framing studies (Hajer and Wagenaar 2003; Hajer 2005).

In the collection of data for the analysis, a three-step approach was followed. First, a literature review of policy and institutional developments in Dutch and US flood governance over the course of the 20th to 21st century was made in order to grasp the major developments in these policy fields. Based on this review, certain periods in the evolution of Dutch and US flood governance were selected for further analysis. These periods marked an important formative moment in the development or change of policy institutions in Dutch and US flood governance.

As a second step, the policymaking process was reconstructed for these periods to trace back the influence of the arguments of experts and political actors on the policy frames co-produced in Dutch and US flood governance. For these reconstructions, transcripts of parliamentary debates and committee hearings were used as a primary source of data. This data was supplemented with secondary materials, such as reports and statements from key actors involved in policymaking processes or the monitoring of these processes (e.g., watchdog organizations like the US Governmental Accountability Office (GAO) or research organizations focused on flood governance).

The overall conclusions of the case study analyses were checked and fine-tuned through stakeholder interviews and discussion meetings in the third step. The aim and setup of these interviews were somewhat different in both case studies. In the US, the interviews were used to gain more insight into the contextual background characteristics at play in a recent case of policymaking on floods (laid out in Chap. 6). Whereas the analysis of congressional records provided much insight into the internal dynamics of this process, it was difficult to get a good overview of the external developments that influenced this process being based in the Netherlands. The interviews held with stakeholders and actors who closely followed this policymaking process helped to better understand the influence of external factors.

In the Netherlands, external factors were easier to grasp. Here, a difficulty rather lay in capturing the full scope of the policymaking process. In contrast to the US, where legislative actions are publicly documented (connected to the congressional committees in which this action takes place), in the Netherlands the process through which laws are formulated and amended takes place behind closed doors. For the Dutch case study, these aspects of the policymaking process, and the influence of experts on this process, were reconstructed based on an analysis of (policy) documents issued by relevant ministries and other governmental bodies in Dutch flood governance, in combination with the parliamentary records that discussed (or not discussed) these documents. To check the findings, discussion meetings were set up with two key Dutch governmental bodies (the Ministry of Environment and Rijkswaterstaat) whose involvement in the policymaking process is not directly documented in parliamentary records.

Unfortunately, the interviews could not be used to check the findings of historical analyses of policymaking processes in Dutch and US flood governance (laid out in Chaps. 3 and 5). However, some valuable data sources were available that well-documented the external factors at play in US flood governance and the full policymaking process in Dutch flood governance. In the US, the books of Arnold (1998) and Barry (1997) were very insightful in this respect (see Chap. 5). In the Netherlands, the energetic documentations of the Zuiderzee Society provided much insight (see Chap. 3). Table 1.1 provides an overview of the data sources used in this book.

Table 1.1 Overview of data sources used in the research

Data collection step	Data sources NL case	Data sources US case
1. Literature review	Review of the scientific and policy literature on the evolution of Dutch flood governance in the 20th–21st century	Review of the scientific and policy literature on the evolution of US flood governance in the 20th–21st century
2. Reconstruction of policymaking process in selected periods	Primary sources: Parliamentary proceedings and committee reports. Secondary sources: Policy reports, scientific reports, policy statements of key stakeholders	Primary sources: Congressional records of House and Senate meetings, committee action (hearings, legislative mark-up sessions and reports). Secondary sources: Policy reports, scientific reports, policy statements of key stakeholders
3. Interviews and discussion meetings	Discussion meeting at the Dutch Ministry of Infrastructure and the Environment with representatives from the flood safety and policy and management departments in October 2015; Discussion meeting at Rijkswaterstaat's Water, Transport and Living Environment Department in January 2016	Interviews with the Federal Emergency Management Agency (FEMA), National Association of Realtors, Research for Institute Resources for the Future, the American Association of State Floodplain Managers, US Government Accountability Office (GAO), National Association of Mutual Insurance Companies in April 2014

1.5 Structure of the Book

The book is built up of six chapters. Chapters 2–5 describe the empirical findings. There are two chapters for each case study. The case study chapters zoom in on specific time periods that either characterize important moments in the institutionalization of the safety approach to floods, or mark an important moment in the shift to spatial measures.

Chapters 2–3 deal with the Dutch case study. Chapter 2 starts out with an analysis of the policymaking process underlying the construction of the Zuiderzee Works (1890–1932), which represents a formative moment in the development of the Dutch safety approach to floods. In Chap. 3, the policymaking process underlying the implementation of three key spatial planning policies in Dutch flood governance are analyzed: the Room for the River project, attempts to implement flood insurance, and the Second Delta Program.

The results of the US case study are presented in Chaps. 4 and 5. Chapter 4 begins with an analysis of the shift to spatial measures in US flood governance, which roughly covers a time period from 1900 to 1960, and proceeds with a description of the institutionalization of the spatial planning approach from the 1970s onwards. Chapter 5 looks at a recent case of policymaking in US flood governance. It studies the process leading up to recent reforms of the US spatial planning policy in 2012 and 2014, which were instigated by the occurrence of hurricane Katrina in 2005.

The book ends with a concluding chapter in which the case study findings are comparatively analyzed. It reflects on the distributive implications of the shift to spatial measures in both case study countries and draws conclusions on the process through which these distributions were produced.

References

Aakre S, Banaszak I, Mechler R (2010) Financial adaptation to disaster risk in the European Union: identifying roles for the public sector. Mitig Adapt Strat Glob Change 15:721–736

Adam B, Beck U, Van Loon J (eds) (2000) The risk society and beyond—critical issues for social theory. Sage Publications, London

Ankersmit FR (1999) Macht door representatie. Kampen, Kok Agora

Arts B, Van Tatenhove J (2004) Policy and power: a conceptual framework between the 'old' and 'new' policy idioms. Policy Sci 37(3/4):339–356

Arts B, Leroy P, Van Tatenhove J (2006) Political modernisation and policy arrangements: a framework for understanding environmental policy change. Publ Organ Rev 6(2):93–106

Arnold JL (1988) The evolution of the 1936 flood control act. Office of History of the United States Army Corps of Engineers, Fort Belvoir

Bachrach P, Baratz MS (1962) Two faces of power. Am Polit Sci Rev 56(04):947–952

Bäckstrand K, Kahn J, Kronsell A, Lövbrand E (eds) (2010) Environmental politics and deliberative democracy—examining the promise of new modes of governance. Edward Elgar Publishing Ltd, Cheltenham

Bader V (2014) Sciences, politics, and associative democracy: democratizing science and expertizing democracy. Innov Eur J Soc Sci Res 27(4):420–441

Barber BR (1984) Strong democracy: participatory politics for a new age. University of California Press, Berkeley, CA

Barry J (1997) Rising tide: the great mississippi flood of 1927 and how it changed America. Simon & Schuster, New York

Beck U (1992) Risk society—towards a new modernity. SAGE publications, London

Beck U (1995) Ecological politics in an age of risk. Polity Press, Cambridge

Benford RD, Snow DA (2000) Framing processes and social movements: an overview and assessment. Ann Rev Sociol 26:611–639

Bergsma E, Gupta J, Jong P (2012) Does individual responsibility increase the adaptive capacity of society? The case of local water management in the Netherlands. J Resour Conserv Recycl 64:13–22

Birkland TA (2007) Lessons of disaster: policy change after catastrophic events. Georgetown University Press, Washington, DC

Bock S (2014) Politicized expertise—an analysis of the political dimensions of consultants' policy recommendations to developing countries with a case study of McKinsey's advice on REDD+ policies. Innov Eur J Soc Sci Res 27(4):379–397

Borowiak CT (2011) Accountability and democracy: the pitfalls and promise of popular control. Oxford Scholarship Online. http://doi.org/10.1093/acprof:oso/9780199778256.001.0001

Botzen WJW, van den Bergh JCJM (2008) Insurance against climate change and flooding in the Netherlands: present, future, and comparison with other countries. Risk Anal 28(2):413–426

Bouwer LM (2010) Disasters and climate change: analyses and methods for projecting future losses from extreme weather. Dissertation, VU University Amsterdam, the Netherlands

Bovens M, Wille A (2011) Diplomatiedemocratie: over de spanning tussen meritocratie en democratie. Bert Bakker, Amsterdam

Bovens M, 't Hart P, Van Twist MJW (2012) Openbaar Bestuur: beleid, organisatie en politiek. Wolters Kluwer Nederland, Alphen aan de Rijn

Bubeck P, Botzen WJW, Aerts JCJH (2012) A review of risk perceptions and other factors that influence flood mitigation behavior. Risk Anal 32(9):1481–1495

Butler C, Pidgeon N (2011) From "flood defence" to "flood risk management": exploring governance, responsibility, and blame. Environ Plan C Govern Policy 29(3):533–547

Clemens E, Cook J (1999) Politics and institutionalism: explaining durability and change. Ann Rev Sociol 25:441–466

Collier SJ (2008) Enacting catastrophe: preparedness, insurance, budgetary rationalization. Econ Soc 37(2):224–250

Cutler AC (2010) The legitimacy of private transnational governance: experts and the transnational market for force. Socio-Econ Rev 8:157–185

Dahl RA (1961) Who governs?: democracy and power in an American City. Yale University Press, New Heaven & London

Dahl RA (1998) On democracy. Yale University Press, New Heaven & London

Dahl RA (2006) On political equality. Yale University Press, New Heaven & London

De Bont R, Vanpaemel G (2012) The scientist as activist: biology and the nature protection movement, 1900–1950. Environ Hist 18(2):203–208

De Marchi, B. (2015). Risk Governance and the Integration of Different Types of Knowledge. In: Fra. Paleo U (ed) Risk governance—the articulation of hazard, politics and ecology. Springer, Dordrecht, pp 149–165

De Swaan A (2004) Zorg en de staat. Welzijn, onderwijs en gezondheidszorg in Europa en de Verenigde Staten in de nieuwe tijd. Amsterdam: Bert Bakker

De Vries J, Wolsink M (2009) Making space for water: spatial planning and water management in the Netherlands. In: Davoudi S, Crawford J, Mehmood A (eds) Planning for climate change—strategies for mitigation and adaptation for spatial planners. Earthscan, London, pp. 191–204)

Dellink R, Den Elzen M, Aiking H, Bergsma E, Berkhout F, Dekker T, Gupta J (2009) Sharing the burden of financing adaptation to climate change. Glob Environ Change 19(4):411–421

Dewulf A, Bouwen R (2012) Issue framing in conversations for change: discursive interaction strategies for "doing differences". J Appl Behav Sci 48(2):168–193

Dewulf A, Craps M, Dercon G (2004) How issues get framed and reframed when different communities meet: a multi-level analysis of a collaborative soil conservation initiative in the Ecuadorian Andes. J Commun Appl Soc Psychol 14(3):177–192

Dewulf A, François G, Pahl-Wostl C, Taillieu T (2007) A framing approach to cross-disciplinary research collaboration: experiences from a large-scale research project on adaptive water management. Ecol Soc 12(2):14. http://www.ecologyandsociety.org/vol12/iss2/art14/. Accessed 24 June, 2015

Dewulf A, Gray B, Putnam L, Lewicki R, Aarts MNC, Bouwen R, van Woerkum CMJ (2009) Disentangling approaches to framing in conflict and negotiation research: a meta-paradigmatic perspective. Hum Relat 62(2):155–193

Dijstelbloem, HO (2007) De democratie anders. Politieke vernieuwing volgens Dewey en Latour. Dissertation, University of Amsterdam, The Netherlands

DiMaggio PJ, Powell WW (1983) The iron cage revisited: institutional isomorphism and collective rationality in organizational fields. Am Sociol Rev 48(2):147–160

Doorn N (2016) Distributing responsibilities for safety from flooding. Available via E3S Web of Conferences. http://doi.org/10.1051/e3sconf/20160724002. Accessed 2 Jun 2018

Dunlop CA (2010) The temporal dimension of knowledge and the limits of policy appraisal: Biofuels policy in the UK. Policy Sci 43(4):343–363

Dunlop CA (2014) The possible experts: how epistemic communities negotiate barriers to knowledge use in ecosystems services policy. Environ Plan C Govern Policy 32(2):208–228
Elgert L (2013) Hard facts and software: the co-production of indicators in a land-use planning model. Environ Values 22(6):765–786
Engelen ER (2008) Op zoek naar legitimiteit: is "disintermediatie" de oplossing voor de "crisis van de democratie?". In: Noorlander CW, Broeksteeg JLW, Hoogers H, Noorlander C (eds) Het volk regeert: beschouwingen over de (Nederlandse) democratie in de 21ste eeuw. Wolf Legal Publishers, Nijmegen, pp 101–126
Entman RM (1993) Framing: towards clarification of a fractured paradigm. J Commun 43(4):51–58
Estlund DM (2008) Democratic authority: a philosophical framework. Princeton University Press, Princeton
Fennema M (1982) Tussen vierde en vijfde macht. In: Fennema M, van der Wouden R (eds) Het Politicologen-debat—Wat is politiek?. Van Gennep, Amsterdam, pp 17–32
Fischer F (1995) Evaluating public policy. Nelson-Hall, Chicago
Flyvbjerg B (2006) Five misunderstandings about case study research. Qual Inq 12(2):219–245
Fournier M, Larrue C, Alexander M, Hegger D, Bakker M, Pettersson M, Crabbé A, Mees H, Chorynski A (2016) Flood risk mitigation in Europe: how far away are we from the aspired forms of adaptive governance? Ecol Soc 21(4):49. https://doi.org/10.5751/ES-08991-210449
Gamson WA, Modigliani A (1989) Media discourse and public opinion on nuclear power: a constructionist approach. Am J Sociol 95(1):1–37
Giddens A (1984) The constitution of society. Polity Press, Cambridge
Giddens A (1999) Risk and responsibility. Modern Law Rev 62(1):1–10
Goffman E (1974) Frame analysis: an essay on the organization of experience. Harvard University Press, Cambridge, MA
Goudie A (2000) The human impact on the natural environment, 5th edn. Blackwell Publishing, Malden
Gralepois M, Larrue C, Wiering M, Crabbé A, Tapsell S, Mees H, Ek K, Szwed M (2016) Is flood defense changing in nature? Shifts in the flood defense strategy in six European countries. Ecol Soc 21(4):37. https://doi.org/10.5751/ES-08907-210437
Gray B, Purdy J, Ansari J (2015) From interactions to institutions: Micro-processes of framing and mechanisms for the structuring of institutional fields. Acad Manag Rev 40(1):115–143
Gusfield JR (1981) The culture of public problems: drinking-driving and the symbolic order. The University of Chicago press, Chicago & London
Haas PM (1992) Introduction: epistemic communities and international policy coordination. Int Organ 46(1):1–35
Habermas J (1996) Between facts and norms—contributions to a discourse theory of law and democracy (trans: Regh W). The Massachusetts Institute of Technology Press, Cambridge MA
Hajer M (1995) The politics of environmental discourse—ecological modernization and the policy process. Oxford University Press, New York
Hajer M (2005) Coalitions, practices, and meaning in environmental politics: from acid rain to BSE. In: Howarth D, Torfing J (eds) Discourse theory in European politics. Palgrave Macmillan, Basingstoke, pp 297–315
Hajer M, Laws D (2008) Ordering through discourse. In: Goodin RE, Moran M, Rein M (eds) The Oxford handbook of public policy. Oxford University Press, Oxford, pp 251–268
Hajer M, Wagenaar H (eds) (2003) Deliberative policy analysis understanding governance in the network society. Cambridge University Press, Cambridge
Hall P (1993) Policy paradigms, social learning, and the state—the case of economic policymaking in Britain. Comp Politics 25(3):275–296
Hall P, Taylor R (1996) Political science and the three new institutionalisms. Polit Stud, XLIV, pp 936–957
Hartmann T (2009) Clumsy floodplains and the law: towards a responsive land policy for extreme floods. Built Environ 35(4):531–544

Hegger DLT, Driessen PPJ, Wiering M, Van Rijswick HFMW, Kundzewiczm ZW, Matczak P, Crabbé A, Raadgever GT, Bakker MHN, Priest SJ, Larrue C, Ek K (2016) Toward more flood resilience: is a diversification of flood risk management strategies the way forward? Ecol Soc 21(4):52. https://doi.org/10.5751/ES-08854-210452

Hegger D, Mees H, Driessen PPJ, Runhaar H (2017) The roles of residents in climate adaptation: a systematic review in the case of The Netherlands. Environ Policy Govern 27(4):336–350

Hindmoor A (2010) Rational Choice. In: March D, Stoker G (eds) Theory and methods in political science, 3rd edn. Palgrave Macmillan, Basingstoke, pp 42–59

Hisschemöller M, Hoppe R (2001) Coping with intractable controversies: the case for problem structuring in policy design and analysis. In: Hisschemöller M, Hoppe R, Dunn W, Ravetz J (eds) Knowledge, power and participation in environmental policy analysis. Transaction Publishers, New Brunswick/London, pp 47–72

Hodgson GM (2007) Institutions and individuals: interaction and evolution. Organ Stud 28(1):95–116

Hoppe R (1999) Policy analysis, science and politics: from speaking truth to power to making sense together. Sci Publ Policy 26(3):201–210

Hoppe R, Petersen A (eds) (1993) Handling frozen fire—political culture and risk management. Westview Press, Boulder/Oxford

Hupe PL, Hill MJ (2006) The three action levels Governance: re-framing the policy process beyond the stages model. In: Peters BG, Pierre J (eds) Handbook of public policy. Sage, London, pp. 13–30

Jasanoff S (1990) The fifth branch: science advisors as policymakers. Harvard University Press, Cambridge, MA

Jasanoff S (ed) (2004) States of knowledge: the co-production of science and social order. Routledge, London

Javeline D, Shufeldt D (2014) Scientific opinion in policymaking: the case of climate change adaptation. Policy Sci 47(2):121–139

Jennings ET Jr, Hall JL (2012) Evidence-based practice and the use of information in state agency decision making. J Public Adm Res Theor 22(2):245–266

Johnson CL, Priest SJ (2008) Flood risk management in England: a changing landscape of risk responsibility? Int J Water Resour Dev 24(4):513–525

Jong P, Van den Brink M (2013) Between tradition and innovation: developing flood risk management plans in the Netherlands. J Flood Risk Manag. http://doi.org/10.1111/jfr3.12070

Jongman B, Koks EE, Husby TG, Ward PJ (2014) Increasing flood exposure in the Netherlands: implications for risk financing. Hydrol Earth Syst Sci 14:1245–1255

Kahan DM (2006) Fear of Democracy: A Cultural Evaluation of Sunstein on Risk. Faculty Scholarship Series, Paper 104: 1071–1109. Available via SSRN. https://papers.ssrn.com/sol3/papers.cfm?abstract_id=801964. Accessed 3 Jun 2018

Kane S, Shorgen JF (2000) Linking adaptation and mitigation in climate change policy. Clim Change 45:75–102

Keessen A, Vink MJ, Wiering M, Boezeman D, Ernst W, Mees H, Van Broekhoven S, Van Eerd M (2016) Solidarity in water management. Ecol Soc 21(4):35

Kingdon J (1995) Agendas, alternatives and public policies. Harper Collins, New York

Knight FH (1921) Risk, uncertainty, and profit. Hart, Schaffner & Marx/Houghton Boston: Mifflin Company

Lasswell HD (1936) Politics: who gets what, when, how. Whitlesey House, New York

Liao KH (2012) A theory on urban resilience to floods: a basis for alternative and intellectual planning practices. Ecol Soc 17(4):48–62

Lijphart A (1999) Comparative politics and the comparative method. Am Polit Sci Rev 65(3):682–693

Low N, Gleeson B (1998) Situating justice in the environment: the case of BHP at the Ok Tedi copper mine. Antipode 30(3):201–226

Lowndes V (2010) The institutional approach. In: March D, Stoker G (eds) Theory and methods in political science, 3rd edn. Palgrave Macmillan, Basingstoke, pp 60–79

Lukes S (2005) Power—a radical view, 2nd edn. Palgrave Macmillan, Hamshire & New York

Lundin M, Öberg P (2014) Expert knowledge use and deliberation in local policy making. Policy Sci 47(1):25–49

Maasen S, Weingart P (eds) (2005) Democratization of expertise? Exploring new forms of scientific advice in political decision-making. Kluwer Academic Publishers, Dordrecht

Mahoney J, Thelen K (2010) A theory of gradual institutional change. In: Mahoney J, Thelen K (eds) Explaining institutional change: ambiguity, agency and power. Cambridge University Press, New York, pp 1–37

Majone G (1999) The regulatory state and its legitimacy problems. West Eur Polit 22(1):1–24

Manin B (2007) The principles of representative government. Cambridge University Press, Cambridge

March J, Olsen J (1989) Rediscovering institutions. Free Press, New York, NY

Maussen, M (2009) Constructing Mosques: the governance of Islam in France and the Netherlands. Dissertation, University of Amsterdam, The Netherlands

Meijerink S (2005) Understanding policy stability and change. The interplay of advocacy coalitions and epistemic communities, windows of opportunity, and Dutch coastal flooding policy 1945–2003. J Eur Publ Policy 12(6):1060–1077

Meijerink S, Dicke W (2008) Shifts in the public-private divide in flood management. Water Resour Dev 24(4):499–512

Merz B, Hall J, Disse M, Schumann A (2010) Fluvial flood risk management in a changing world. Nat Hazards Earth Syst Sci 10(3):509–527

Metze T (2014) Fracking the debate: frame shifts and boundary Work in Dutch decision making on shale gas. J Environ Plan Policy Manage 19(1):35–52

Mill JS (1861) Considerations on representative government, 1st edn. Parker, Son & Bourn, London

Neuvel JMM, Van der Knaap W (2010) A spatial planning perspective for measures concerning flood risk management. Water Resour Dev 26(2):283–296

North DC (1990) Institutions, institutional change and economic performance. University Press, Cambridge

Nowotny H, Scott P, Gibbons M (2001) Re-thinking science: knowledge and the public in an age of uncertainty. Polity Press, London

Paavola J, Adger WN (2002) Fair adaptation to climate change. Ecol Econ 56:594–609

Pahl-Wostl C (2007) Transitions towards adaptive management of water facing climate and global change. Water Resour Manage 21(1):49–62

Pahl-Wostl C, Jeffrey P, Isendahl N, Brugnach M (2011) Maturing the new water management paradigm: progressing from aspiration to practice. Water Resour Manage 25(3):837–856

Parsons C (2012) Constructivism and interpretive theory. In: March D, Stoker G (eds) Theory and methods in political science, 3rd edn. Palgrave Macmillan, Basingstoke, pp 80–98

Paudel Y, Botzen WJW, Aerts JCJH, Dijkstra TK (2015) Risk allocation in a public-private catastrophe insurance system: an actuarial analysis of deductibles, stop-loss, and premiums. J Flood Risk Manag 8(2):116–134

Penning-Rowsell EC, Pardoe J (2012) Who benefits and who loses from flood risk reduction? Environ Plan C Govern Policy 30(3):448–466

Rawls J (1999) A theory of justice, revised edition. Oxford University Press, Oxford

Rayner S (2012) Uncomfortable knowledge: the social construction of ignorance in science and environmental policy discourses. Econ Soc 41(1):107–125

Reddy SG (1996) Claims to expert knowledge and the subversion of democracy: the triumph of risk over uncertainty. Econ Soc 25(2):222–254

Rein M, Schön D (1993) Reframing Policy Discourse. In: Fischer F, Forester J (eds) The argumentative turn in policy analysis and planning. Duke University Press, Durham, pp 145–166

Renn O (1992) Concepts of risk: a classification. In: Krimsky S, Golding D (eds) Social theories of risk. Praeger Publishers, Westport, pp 53–79

Renn O, Schweitzer P (2009) Inclusive risk governance: concepts and application to environmental policy making. Environ Policy Govern 19:174–185

Rietig K (2014) 'Neutral' experts? How input of scientific expertise matters in international environmental negotiations. Policy Sci 47(2):141–160

Rijke J, Van Herk S, Zevenbergen C, Ashley R (2012) Room for the river: delivering integrated river basin management in the Netherlands. Int J River Basin Manag 10(4):369–382

Rimkutė D, Haverland M (2015) How does the European Commission use scientific expertise? Results from a survey of scientific members of the Commission's expert committees. Comp Eur Polit 13(4):430–449

Rittel HWJ, Webber MM (1973) Dilemmas in a general theory of planning. Policy Sci 4(2):155–169

Rosanvallon P (2008). Counter-democracy—politics in an age of distrust (trans Goldhammer A). Cambridge University Press, Cambridge

Sabatier PA, Jenkins-Smith (1988) An advocacy coalition framework of policy change and the role of policy-oriented learning therein. Policy Sciences 21 (2–3):129–168

Sayers PB, Hall JW, Meadowcroft IC (2002) Towards risk-based flood hazard management in the UK. Civ Eng 150(5):36–42

Scott JC (1998) Seeing like a state. How certain schemes to improve the human condition have failed. Yale University Press, New Haven & London

Seawright J, Gerring J (2008) Case selection techniques in case study research: a menu of qualitative and quantitative options. Polit Res Q 61(2):294–308

Shue H (1999) Global environment and international inequality. Int Aff 75(3):531–545

Skocpol T (1979) States and social revolutions: a comparative analysis of France, Russia, and China. Cambridge University Press, Canto Classics. Cambridge

Steinberg MW (1998) Tilting the frame: considerations on collective action framing from a discursive turn. Theory Soc 27(6):845–872

Swiss Re (2012) Flood—an underestimated risk: inspect, inform, insure. Zurich, Swiss Reinsurance Company Ltd., Zurich. Available via Swiss Re. http://www.swissre.com/library/Flood__an_underestimated_risk_Inspect_inform_insure.html. Accessed 3 Jun 2018

Turner S (2001) What is the problem with experts? Soc Stud Sci 31(1):123–149

UNISDR (2015) Disasters in Numbers. United Nations Office for Disaster Risk Reduction: Geneva. Available via UNISDR. http://www.unisdr.org/files/47804_2015disastertrendsinfographic.pdf. Accessed 3 Jun 2018

Van Buuren A, Klijn E, Edelenbos J (2012) Democratic legitimacy of new forms of water management in the Netherlands. Int J Water Resour Dev 28(4):629–645

Van der Steen MA, Van Twist MJM (2013) Foresight and long-term policy-making: an analysis of anticipatory boundary work in policy organizations in The Netherlands. Futures 54:33–42

Van Gunsteren H (1976) The quest for control: a critique of the rational-central-rule approach in public affairs. Wiley, New York

Van Hulst M, Yanow D (2016) From policy "frames" to "framing": theorizing a more dynamic, political approach. Am Rev Publ Admin 46(1):92–112

Van Tatenhove J, Arts B, Leroy P (eds) (2000) Political modernisation and the environment: the renewal of environmental policy arrangements. Kluwer Academic Publishers, Dordrecht, Boston & London, pp 53–69

Verloo M (2005) Mainstreaming gender equality in Europe. A critical frame analysis approach. Greek Rev Soc Res 117(B): 11–34

Vogel KM (2008) 'Iraqi Winnebagos™ of death': imagined and realized futures of US bioweapons threat assessments. Sci Publ Policy 35(8):561–573

Warren ME (2006) Deliberative democracy and authority. Am Polit Sci Rev 90(1):46–60

Waylen K, Young J (2014) Expectations and experiences of diverse forms of knowledge use: the case of the UK national ecosystem assessment. Environ Plan C Govern Policy 32(2):229–246

Weber M (1978) Economy and society (trans and ed Roth G, Wittich C). University of California Press, Berkeley & Los Angeles

Weiss CH (1979) The many meanings of research utilization. Public Adm Rev 39(5):426–431

Wesselink A, Buchanan KS, Georgiadou Y, Turnhout E (2012) Technical knowledge, discursive spaces and politics at the science–policy interface. Environ Sci Policy 30(1):1–9

Wiering MA, Arts BJM (2006) Discursive shifts in Dutch river management: 'deep' institutional change or adaptation strategy? Hydrobiologia 565(1):327–338

Wiering MA, Immink I (2006) When water management meets spatial planning: a policy-arrangements perspective. Environ Plan C Govern Policy 24(3):423–438

Wiering M, Kaufmann M, Mees H, Schellenberer T, Ganzevoort W, Hegger D, Larrue C, Matczak P (2017) Varieties of flood risk governance in Europe: how do countries respond to driving forces and what explains institutional change? Glob Environ Change 44:15–26

Winsemius HC, Van Beek LPH, Jongman B, Ward PJ, Bouwman A (2013) A framework for global river flood risk assessments. Hydrol Earth Syst Sci 17:1871–1892

Winsemius HC, Jongman B, Veldkamp TIE, Hallegatte S, Bangalore M, Ward PJ (2018) Disaster risk, climate change, and poverty: assessing the global exposure of poor people to floods and droughts. Environ Dev Econ. http://doi.org/10.1017/S1355770X17000444

Wolsink M (2006) River basin approach and integrated water management: governance pitfalls for the Dutch space-water-adjustment management principle. Geoforum 37(4):473–587

World Economic Forum (2015) Global Risks 2015, 10th edn. World Economic Forum, Geneva. Available via World Economic Forum. http://reports.weforum.org/global-risks-2015/. Accessed 3 Jun 2018

Yin RK (2009) Doing case study research, 4th edn. Sage, Thousand Oaks

Zanetti LA, Adams GB (2000) In service of the Leviathan: democracy, ethics and the potential for administrative evil in the new public management. Admin Theory Praxis 22(3):534–554

Chapter 2
Establishing Safety Institutions in Dutch Flood Governance: A Political Genealogy of the Zuiderzee Works

2.1 Introduction

Delta's—areas where rivers mouth into seas—belong to the world's most densely populated areas (Van Urk et al. 1999). They are located conveniently for sea trade and their floodplains provide highly fertile plots of land. However, deltas are also very vulnerable to flood risks. Events such as hurricane Katrina in the US in 2005, the 2011 floods in the Vietnamese Mekong Delta, and the recent Philippine floods demonstrate this vulnerability. Growing awareness of climate change leads delta-countries to look across their own borders for finding ways to better adapt to the flood risks they face.

The Netherlands deserve particular attention in this respect. The Dutch are famous for their Afsluitdijk—a 32 km long dam that closes off an inner sea (the Zuiderzee) and protects the area behind it. And there are the impressive Delta Works, a system of dykes that protects the south-western part of the Netherlands. Such large-scale protection works are not only difficult to implement technologically but also politically. Devised by experts to ward off the risk of flooding, their implementation often has negative implications for other groups in society and can therefore conflict with other interests like those of trade, fishery, and the environment. These interests have to be weighed against the goal of providing protection against uncertain future flood risks. Policymakers rely heavily on expert-knowledge in making such trade-offs. The risk prognoses made by these experts reduce some of the uncertainty surrounding future risk of flooding and as such help policymakers act in indeterminate circumstances. However, this reliance on expert-knowledge has raised concerns about the "political" trade-offs made in risk governance; the fear generally is that value conflicts and conflicts of interest are insufficiently recognized under their "technical" or "operational"

This chapter is co-authored by prof. M. Fennema, Department of Political Sciences, University of Amsterdam, m.fennema@uva.nl.

E. Bergsma, *From Flood Safety to Spatial Management*, Water Governance - Concepts, Methods, and Practice, https://doi.org/10.1007/978-3-319-96716-5_2

understanding of risks (Beck 1992; Habermas 1996; Rosanvallon 2008; Vogel 2008; Rayner 2012).

In order to learn from Dutch experiences, this chapter reconstructs the policy-making process of the Zuiderzee Works, a process that roughly covers the period 1880–1932. This period marks an important moment in the formation of the Dutch safety approach to floods. This chapter analyzes the role of experts in this formative process through a framing perspective. How have experts, through their discursive interaction with policymakers, influenced the policy discourse on floods and, through this, the distributive aspects of the Zuiderzee Works policy? The policy reconstruction is based on an analysis of parliamentary records, historical (policy) documents, and additional scientific and policy literature.

The chapter is structured as follows. Section 2.2 sketches the political context in which a new group of experts emerged who grew devoted to the plan to close off the Zuiderzee. Section 2.3 analyses how two key concerns that were raised against the Zuiderzee Works were addressed in the policymaking process. Section 2.4 shortly reflects on the implementation of the Zuiderzee Works over the years. Section 2.5 reflects on the role of experts in the formation of the Dutch safety approach and discusses its implications for distributive decision-making under this approach.

2.2 From Plan to Policy

2.2.1 The Rise of a Progressive-Liberal Elite

Both the emergence of the plan to close off the Zuiderzee in the second half of the 19th century and its rising political salience in the 20th century can best be explained in the particular political context of the Netherlands in this period. This context created a window of opportunity for a new intellectual elite that was firmly committed to the realization of the Zuiderzee Works.

In 1848, a hallmark year in Dutch constitutional development, a major constitutional reform was passed under the leadership of liberal politician Rudolf Thorbecke. The new constitution was based on a liberal doctrine that advocated minimal state influence to reduce the abuse of power by the state (Drentje 2011). However, over the years, the agriculture-based economy of the Netherlands started to lag behind neighboring countries that benefitted from new industrial technologies (Ter Veen 1935). Against the backdrop of aggravating social problems, a new intellectual elite emerged that challenged the conservative-liberal state doctrine.

The attitude of this elite has been described by Baneke (2011: 106) as “synthetic technocratic”: They rejected specialization, short-term materialism, and narrow individual or corporate interests in favor of the technocratic ideal of politics. Sharing an urge for meritocratic leadership, this elite was in favor of a government of manufacturers, traders, and bankers who had successfully demonstrated their productive capacities, as well as of the influence of scientists, planners, and civil engineers who were able translate social-technocratic ideals into questions of rational state planning (Den Hoed and Keizer 2007: 72). Van der Vleuten (2004) traces this intellectual movement back to the 18th and early 19th century, when thinkers like Turgot, Condorcet, Saint-Simon, Bentham, James, and John Stuart Mill stressed the importance of improved access to and circulation of rational knowledge and technology for commerce and industry. Like its modern counterparts, the elite held an organic vision on society; the state as the "body politic" should create the right conditions to maximize the productive potential of society. A key instrument at its disposal was the implementation of large-scale infrastructural projects, such as railroads and waterways. Internationally, such ideas found their application for example in the construction of the Suez-Canal (1859–1869).

In the Netherlands, central banker H.P.G. Quack (1834–1917) disseminated the ideas of Henri de Saint-Simon (Quack 1915). Inspired by his writings, members of the upper middle class and aristocracy, including (urban) planners, engineers, bankers, landed gentry, and politicians who sympathized with the technocratic ideology organized themselves in private associations to challenge the liberal state doctrine. Through these associations, they initiated privately funded research into specific social problems, such as health care, schooling, or transportation, to draft plans for governmental action. Engineers, whose profession "socialized" during the second half of the 19th century, played a leading role in these developments (Lintsen 1980). Armed with research reports, these engineer-led associations sought to influence public opinion by writing articles in national newspapers and journals, publishing brochures, and organizing public debates so as to pressure those in government to take action (Van den Brink and Molema 2008). Examples of such associations are the Association for the General Wellbeing,[1] the Central Bureau for Social Advice,[2] and the Association of Democratic Engineers and Architects.[3]

By the end of the 19th century, the dominant Liberal Party split into a conservative and a progressive branch. Conservatives remained attached to a laissez fair policy and rejected general suffrage. Progressive liberals, like Tak van Poortvliet, Cornelis Lely, Hendrikus Colijn, and later Hans Max Hirschfeld (see also Fennema and Rhijnsburger 2007), advocated extensions of the census and were in favor of expanding the role of the state, particularly in the national infrastructure. This significantly contributed to the influence of the new liberal-progressive elite in the Netherlands.

The ideology that became pervasive under the new elite was based on a close harmony of business and state. Progressive liberals in parliament fostered intimate relationships between Dutch state institutions and the private research associations formed by members of the new elite. Under these circumstances, the Dutch state became increasingly involved in the establishment of provisions to improve the national infrastructure, such as railroad construction, electricity supply, and water management. The plan to reclaim the Zuiderzee was one of these Saint-Simonian projects, comparable with the Suez Canal (1869) and the Panama Canal (1914).

[1]Maatschappij tot Nut van 't Algemeen.

[2]Centraal Bureau voor Sociale Adviezen.

[3]Sociaal-Technische Vereeniging van Democratische Ingenieurs en Architecten.

2.2.2 *The Engagement of Progressive Elites in Zuiderzee Reclamations*

The Zuiderzee has always been an important area for the Netherlands. It formed a sea trade passage that contributed to the economic prosperity of the Netherlands in the 16th and 17th centuries, it has been an important building block of Dutch military defense works,[4] and it provided rich fishing grounds. However, it also posed a threat to the people living and working near its shores; the Zuiderzee has been known for its devastating storm surges, flooding whole islands and villages and killing thousands of people (Braat 1932).

The first plan to reclaim the Zuiderzee dates back to 1667, when the son of the renowned mathematician and engineer Simon Stevin, Hendrik Stevin, pondered about closing off and reclaiming the Zuiderzee. No one really took his ideas seriously at that time; reclaiming a whole sea was considered technically impossible. However, the successful reclamation of the large lake De Haarlemmermeer in 1852 prompted a renewed interest in Zuiderzee reclamation. The reclamation of the Zuiderzee dovetailed well with the utopian visions like that of Robert Owen and Charles Fourrier, as it foresaw the setting up of farmers' communities in a new and empty land. As such, it attracted members of the newly emerging liberal-progressive elite, many of who also saw possibilities for communal experiments in the colonies.

Two other factors contributed to the interest of the progressive elite in Zuiderzee reclamation. First, Dutch coastal water management was still in its infancy. While a central water management authority (Rijkswaterstaat) had been established by the French administration in 1798, the work of this organization focused mainly on the inland water system. Coastal water management always lagged behind, also because there was not much expertise in this area. Up to the 18th century, the sea had been primarily studied by cartography for optimizing trading routes (Toussaint 2009). However, the work of the French mathematician Laplace on ocean tides had inspired a new science on the dynamics behind tidal flows, which also increased opportunities for control in this area. The new elite, among who were many hydraulic engineers, thus saw in Zuiderzee reclamation an opportunity to put their expertise to use in this until then under-researched policy domain of flood safety.

Second, up to the 19th century, land reclamation activity was largely an unplanned endeavor (Danner 1992); reclamation was a business activity, where the Dutch government used private capital to drain fruitful areas, which were then leased to investors who hoped to return their investments with the profits gained by cultivating the area. This business often left farmers on the newly recovered lands deprived of social and physical provisions like road networks, schools, and churches (Ter Veen 1925). When in 1866 the private Land Banking Corporation[5] requested a concession to reclaim parts of the Zuiderzee, the progressive-liberal elite raised its voice and argued that

[4]It was part of the "Stelling van Amsterdam".

[5]Maatschappij voor Grondkrediet.

such an activity, which so clearly involved matters of common interest, should be undertaken and financed by the state.

2.2.3 The Zuiderzee Society and Its Achievements

Between 1848 and 1875, several plans to reclaim the Zuiderzee had been circulated.[6] The first plan that was adopted by parliament was developed by engineer W.F. Leemans in 1875. While his plan was never implemented, it did draw the attention of Age Buma (1820–1893, see Textbox 2.1), a deputy from Friesland. In 1884, Buma received a letter from his friend in London who had written about a group of investors in the United Kingdom that was willing to invest in Zuiderzee reclamations.[7] Buma, however, was skeptical about the influence of private capital in reclamation activities without a government overseeing the whole enterprise. Together with P.J.G. van Diggelen (1837–1907), the son of an engineer who had proposed to empolder the Zuiderzee as early as 1849, Buma established the Zuiderzee Society in 1886 to instigate a technical and financial study into the possibilities of closing off and later gradually reclaiming the Zuiderzee.

Textbox 2.1 Age Buma (source: http://www.parlement.com/id/vg09lkyz2axz/a_buma)

Age Buma (1820–1893). With a background in agriculture, Buma was elected representative of the Sneek district (Friesland) in the Lower House from 1882 until 1888 as part of the liberal progressive union that supported Tak van Poortvliet in his efforts to extent the census. Buma was a member of the North-Holland Friesland railway committee and a board member of the Association for Civil Education, but in parliament he acted as a spokesman on water management, in particular on Zuiderzee reclamation plans. From 1886 until his death in 1893, Buma was chairman of the Zuiderzee Society.

[6] An overview of these plans can be found in Jansma (1954).

[7] This letter was written by De Jongh van Arkel, who informed Buma about “een Agent van een groep kapitalisten alhier (is), die niet ongenegen zouden zijn dit groote en wetenschappelijke werk te ondernemen, indien de onderneming, naar Uw gevoelen, de gewenschte voordeelen aan de ondernemers zoude kunnen afwerpen, en de Concessie daarvoor van het Nederlandsche Gouvernement zou kunnen worden verkregen”. (Cleintuar 1982: 41).

Buma and Van Diggelen managed to collect the necessary funds to commission a renowned hydraulic engineer, J. van der Toorn, to conduct the research. Van der Toorn employed a young engineer, Cornelis Lely (1854–1929), with whom he had collaborated at Rijkswaterstaat (Jansma 1954). When Van der Toorn left prematurely because of a financial conflict in 1887, it was Lely who took over Van der Toorn's position and developed a plan to close off and reclaim large parts of the Zuiderzee (see Textbox 2.2).

Textbox 2.2 Cornelis Lely (source: http://www.biografischportaal.nl/persoon/05123326)

Cornelis Lely (1854–1929). Already in high school Lely was noted for his mathematical skills. He successfully completed the prestigious Polytechnic academy in Delft. While not very successful in his career as a hydraulic engineer, he worked his way up in politics, occupying the post of Minister of Water Management three times. Lely's "enlightened stewardship" vision is reflected in his contribution to several social (i.e., on education and healthcare) and economic (i.e., on state mining) laws (Cleintuar 1990: 25).

Lely carefully presented his plan in eight technical notes (Zuiderzeevereniging 1892a). He wanted to recover a maximum amount of hectares (approximately 200,000) of the sea's most fertile grounds in four separate empolderings while also making sure that the surface of the remaining lake would be large enough to absorb rising water levels caused by storms. He calculated the optimum depth of the lake so that its water levels would be high enough to maintain inland waterway transport but still low enough to allow the country's major rivers (especially the IJssel) to discharge into the lake. Besides the advantage of new land to cultivate, Lely emphasized the additional gains that would result from closing off the Zuiderzee with an enclosure dyke (the Afsluitdijk). He argued that not only the chance of flooding would be reduced, but the maintenance costs of polder dykes would also be significantly lower. Agricultural production in the provinces situated around the Zuiderzee that suffered from salinization would benefit from the creation of a large freshwater lake. Lely stressed that these benefits could only be reaped if the project was led by an organization with a focus on the long-term interest of the nation, that is, the Dutch state. He estimated that the government would have to invest 192 million guilders.

When Lely's notes were made public in 1888, Lely and his Zuiderzee Society had to overcome resistance to their plans. Concerns were expressed about the financial attainability of the project and the consequences for the Zuiderzee fishing industry. How did they manage to overcome these concerns and generate the necessary public and political support for the adoption of the Zuiderzee Act in 1918? To answer this question, the next section analyzes the political efforts of Cornelis Lely and the Zuiderzee Society in relation to the key concerns mentioned above.

2.3 The Political Efforts of the Zuiderzee Society

2.3.1 Dealing with Arguments of a Technical and Financial Nature

From the start, the Zuiderzee Society maintained close ties with liberal-progressive politicians. Not only was the association set up by a progressively minded politician (Buma), but its efforts were also supported by key politicians such as J.P.R. Tak van Poortvliet (1839–1904), who in 1877 became Minister of Transportation and Commerce and in 1891 Minister of Home Affairs. Descendant from the landed gentry in Zeeland, Tak's progressive ideals were fuelled by his friend H.P.G. Quack.[8] While Tak is primarily known for his efforts to extend the census, he was an expert on water management and took a special interest in the Zuiderzee Society.[9]

Before his last note on the Zuiderzee Works was published, Lely had been appointed Minister of Water Management in the progressive-liberal government of Tak van Poortvliet/Van Tienhoven (1891–1894). In 1892, the Zuiderzee Society circulated Lely's eight technical notes among the members of Dutch parliament. While in principle well-received by most of his progressively minded colleagues, Lely's plan drew out criticism from conservatives who were worried about the impact on the national treasury as well as from the state's civil engineers at Rijkswaterstaat who doubted the technical feasibility of the plan. These two types of criticisms often went hand in hand; the budgetary calculations underlying Lely's plan were called into question because it was thought that the works would be much more complex and would take much longer than Lely anticipated. Lely, as Minister of Water Management, now faced the difficult task of deciding on his own plans.

Well aware of his delicate position, Lely installed an independent state commission of experts to study his plans. Although a small minority felt that the project was too costly, the majority accepted Lely's conclusion that his plan concurred with the general interest and should be executed by the state (Staatscommissie 1894).

[8]For whom Tak travelled through Germany and Switzerland to look for writings of old revolutionary thinkers in antiquarian bookshops (http://www.parlement.com/id/vg09lla1e0x6/j_p_r_tak_van_poortvliet).

[9]Molhuysen and Blok (1918: 1296) write: "It may be so that nobody contributed more to the improvement of existing (water)ways than Tak.".

However, when the state commission published its advice in 1894, the Tak van Poortvliet government made a proposal to modernize the suffrage system, which led to its fall (Van der Ham 2007; Jansma 1954). The conservative-liberal government that was subsequently formed was hesitant to spend large amounts of money on such a high-risk project.

When a progressive government (the Pierson government) formed again in 1897, Lely was reinstalled as Minister of Water Management. It was left to Lely to convince his fellow engineers, many of who worked within the ranks of the central water managed authority Rijkswaterstaat. These Rijkswaterstaat engineers were critical toward Lely's plans, resulting from a broader conflict of visions on the role of engineers in society between "military" Rijkswaterstaat engineers for whom the engineering profession was a pure technical vocation and "civil" engineers active in private research associations who pursued an extended role of engineers in social affairs (Lintsen 1980: 243–299). When Lely presented his notes, Rijkswaterstaat engineers deemed the Zuiderzee project a too-risky business for the state to undertake. Lely visited Rijkswaterstaat many times to convince its members of the technical and financial attainability of his plans.

Textbox 2.3 Hendrik Christiaan van der Houven van Oordt (source: http://www.biografischwoordenboekgelderland.nl/bio/3_Hendrik_Christiaan-_van_der_Houven_van_Oordt)

Hendrik Christiaan van der Houven van Oordt (1837–1901) was an industrialist and landed property owner, a status he owed mostly to his accomplishments on the private land auction market. In 1877, he bought a small lake (Horstermeer), which he reclaimed for cultivation. Besides a member of the Provincial States of Gelderland, he held the position of dike warden. Until his death in 1901, he acted as the secretary of the Zuiderzee Society. His most influential work, "The Economic Benefits of the Zuiderzee Works", which was printed in 1898 by his brother (publisher Brill in Leiden) and sent to all municipal councils, sold 1600 copies in local bookstores in one year.

The Zuiderzee Society started campaigning for Lely's plans in broader political circles. In 1898, the association published a report on the economic benefits of the Zuiderzee Works, written by the association's secretary, Van der Houven van Oordt (see Textbox 2.3), together with a young economist, G. Vissering. The report addressed the benefits of new land for cultivation, better flood protection, and an improved inland water system. For the first time, population growth, which increased relatively fast in the period between 1890 and 1900 (NIDI 2003: 13), was incorporated as an argument; new jobs would be needed to support the growing population, and these jobs could be provided by the Zuiderzee Works. The authors stressed that the financial impacts of the Zuiderzee Works, both positive and negative, would not be felt immediately but would be spread over a considerable period of time (der Houven et al. 1898: 39).

When Lely encountered resistance from the Minister of Finance, he drafted a bill that only included the construction of the two cheapest polders of the four originally included in his plan. Accompanied by revised reports on the economic benefits of the works edited by Vissering (Van der Houven-van Oordt and Vissering 1998, 1901, see Textbox 2.4), this bill was passed in 1901.

Textbox 2.4 Gerard Vissering (source: http://www.nieuwlanderfgoed.nl/archief/waterschrijvers/v)

Gerard Vissering (1865–1937) has been one of the most devoted supporters of the Zuiderzee Works. As the son of one of the nation's leading economist who became minister of Finance and chief editor of De Gids, he quickly made a career in (international) finance. He was director of the Bank of Amsterdam (1900–1906), president of the Java Bank (1906–1912), and president of the Dutch Central Bank (1912–1931), occupations that he combined with his services for the Zuiderzee Society, first as secretary (1901–1906) and later as chairman (1919–1937). Being a prolific and rhetorically strong writer, he published many reports and newspaper articles on the Zuiderzee Works. He was a sailor and a skater and invented a new type of skate: the Vissering-Ruiter model.

The Pierson government did not serve its full term and Lely was replaced by the conservative-liberal De Marez Oyens, who asked two inspector-generals of Rijkswaterstaat to reevaluate Lely's plans. The inspector-generals judged Lely's plans technically feasible but financially unattainable. Lely, however, found a helping hand in the vice president of the Zuiderzee Society, W.F. Leemans, who was also director at Rijkswaterstaat. In an accompanying letter to the report of his subordinates, Leemans argued that "every nation, from time to time, has to initiate great work without fixating on the costs"[10] (Zuiderzeevereniging 1905b: 109).

The Zuiderzee Society increasingly targeted the general public (e.g., through public brochures, the national media). Some of its opponents started using similar strategies. Derk Roelof Mansholt, a landed farmer from the northern province of Groningen, had presented calculations that ran counter to those of the Zuiderzee Society (Krips-Van der Laan 1999). In several newspaper articles and popular brochures, he repeatedly described the plans of the Zuiderzee Society as a costly project aimed to reclaim a "worthless swamp", while the abundant and much cheaper opportunities to improve existing wastelands for cultivation were disregarded (Ter Laan 1949: 13).

When the progressive-liberals assumed power again in 1905, the new prime minister decided not to reinstall Lely as Water Management because the Zuiderzee Works, strongly linked to Lely in person, had become a contentious political issue (Jansma 1954: 138). Instead, J. Kraus was appointed, who was enthusiastic about Lely's plans but as a newcomer hesitated to make major decisions on such a controversial topic. Kraus therefore sent a new bill to parliament in 1907 that aimed at the construction of only one small "test polder" (Wieringenmeer). This bill caused heated debates in parliament, which forced Kraus to commission a series of new studies on different aspects of the plan (Van Blom 1917: 132–133). With the promise that these studies would be awaited before the minister would start constructing the test polder, the bill was passed in 1909.

Kraus's bill was not well received by members of the Zuiderzee Society. However, as two of its most energetic members were abroad—Vissering had been appointed president of the Javasche Bank in Batavia in 1906 and Lely left Holland to become governor of Suriname in 1902—it was now left to men like Harm Smeenge, member of the Lower House, the young engineer Auguste Plate (see Textbox 2.5), and social geographer Anton Beekman to defend Lely's original plans. They failed to change Kraus 's bill.

[10]Original Dutch citation: "een volk moet van tijd tot tijd een groot werk aanvatten en daarbij niet zien op de kosten".

Textbox 2.5 Auguste Plate (source: http://resources.huygens.knaw.nl/bwn1880-2000/lemmata/bwn1/platea)

Auguste Plate (1881–1953) was the son of Antoine Plate, who founded the Holland America Line and who was a great admirer of Ferdinand de Lesseps, who designed and constructed the Suez Canal. He was a member of the Association of Democratic Engineers and Architects (STV). Just before he moved to Indonesia to work for the Nederlandsch-Indische Spoorwegmaatschappij, Plate conducted a study on the effects of the Zuiderzee Works on national unemployment in 1914. After his return in 1917, he became a central figure in urban development in Rotterdam, where he advocated decent housing for the working class and because of his socialist learning was nicknamed Pink Plate.

It was only when Vissering returned to the Netherlands in 1912 that progress was made again. In 1913, Lely was installed as Minister of Water Management for the third time and Queen Wilhelmina addressed the importance of the Zuiderzee Works in her speech to the yearly Assembly of the two Houses. On return to his ministerial post, Lely withdrew Kraus's 1909 bill. Delayed by World War I, Lely did not submit a new Zuiderzee bill to parliament until 1916.

This time circumstances were favorable. The bill was presented just after the Zuiderzee region had been hit by severe floods that killed 20 people and caused huge damages. The floods underlined the importance of the Afsluitdijk for national safety while World War I demonstrated the need for self-sufficiency in agricultural production. Because the international reputation of the Dutch had been dented by its neutral position in the war, people welcomed a project like the Zuiderzee Works that would boost the national confidence by demonstrating the Dutch employed a peaceful land annexation strategy (Van der Geest et al. 2008; Beyen 2008).

The Society launched a final attack on its opponents. It was especially the Mansholt family who required a response. While Derk Roelof Mansholt increasingly used national media to condemn Lely's plans, his son, Lambertus Helbrig, deputy of Groningen, openly worried about Lely's promises to increase the height of coastal dykes in the provinces of Noord-Holland and Friesland to counter rising water levels caused by the construction of the Afsluitdijk while no such provisions were made for Groningen. In their rejoinder, the Zuiderzee Society played the nationalist card:

"[F]ortunately our country does not consist of many such prophets of doom; for we wouldn't have been where we are right now"[11] (Zuiderzeevereeniging 1916: 83).

From March 7 to June 21, 1918, Lely's Zuiderzee bill was discussed in parliament (Zuiderzeevereniging 1920). Lely found an important ally in Zuiderzee Society's vice president Smeenge (see Textbox 2.6). It was especially when Lely's personal involvement was held against him that Smeenge came to Lely's rescue with long interventions on the hard work and many accomplishments of Lely and the Zuiderzee Society. Lely himself primarily dealt with technical concerns and questions.

Textbox 2.6 Harm Smeenge (source: https://www.geni.com/people/Harm-Smeenge/6000000031430738009)

Harm Smeenge (1852–1935) took part in Lely's 1892 state commission, after which he joined the Zuiderzee Society's executive committee in 1897. In 1906 he was appointed vice president, which he combined with his presidency of an association for the inland shipping sector. He occupied a seat in the Lower House from 1888–1919 and a seat in the Upper House from 1920 onward. Until his death in 1935, he passionately voiced the ideas and concerns of the Zuiderzee Society in parliament.

The Mansholt family name was often mentioned in the parliamentary debates when questions were raised as to whether the Afsluitdijk and polder-dikes could indeed be as low as Lely assumed. Lely and Smeenge treated the concerns of Mansholt with some disdain; for them the focus should be on the common good, not on the partisan interests of Groningen. An example of this can be found in the reaction of Lely to raised concerns about flood safety in Groningen: "As I said, the key point of the bill is twofold, it comprises land annexation and improvement of water conditions, both purely matters of the common good, a common good that does not relate to one specific province but to our country at large"[12] (Kamerstukken II 1917/1918a: 1975). However, Lely did make a pledge to install a state commission to investigate the altitude of the Afsluitdijk. This commission was headed by the Nobel

[11] Original Dutch quotation: "Gelukkig dat ons land niet uit velen zoo angstvallige ongeluksprofeten bestaat; het zou dan nimmer geworden zijn wat het nu is.".

[12] Original Dutch quotation: "De hoofdgedachte van het wetsontwerp is, zooals ik reeds zeide, tweeledig, namelijk landaanwinning en verbetering van de waterstaatkundigen toestand, beide zaken van zuiver algemeen belang en in deze wel van een algemeen belang, dat betrekking heeft niet op een enkele provincie, maar op een overgroot deel van ons land.".

Prize winner Prof. H.A. Lorentz (1853–1928), who in 1926, after years of research, advised to elevate the Afsluitdijk at several places (Fuchs and Simons 1972).

Rijkswaterstaat employees had questioned the assumed storage capacity of the IJssel Lake, as members of the Zuiderzee Society had started to call the empoldered Zuiderzee, and similar questions were raised in parliament. Lely countered these arguments with all sorts of technical details, but he did not downplay the remaining uncertainties. Although he emphasized the long-time path of the Works, which allowed for continuous learning and adjustment, he argued that all great works require risk-taking in order to advance social progress.

Apart from these technical and financial concerns, the general attitude in parliament was positive, and this was a major breakthrough compared to earlier parliamentary debates on Zuiderzee bills. In general, the significance of the Works for the nation was acknowledged and Lely's relentless efforts were applauded. The concerns of the Minister of War—who objected to a fast execution of the Works, as they would destroy the nation's most important water defense unit (the Stelling van Amsterdam)—were dismissed. At times, the mood even turned somewhat euphoric. Carried away by their imagination, deputies discussed the shape of the parcels (of which the bill made no reference at all) on the recovered lands; appalled by the previous speaker who liked straight plots better than curved ones, one deputy exclaimed: "But one does not turn a province into a checkerboard!"[13] (Kamerstukken II 1917/1918b: 1888).

Later authors have explained the sudden change in attitude toward the Zuiderzee Works from favorable circumstances like the 1916 floods and the First World War (e.g. Beyen 2008; Fuchs and Simons 1972). However, Lely and the Zuiderzee Society definitely contributed to making these circumstances favorable to their cause. Beekman, for example, writes that "while the flood risk is not the only and definitely not the most important driving force behind the Zuiderzee Works, it speaks more forcefully to the masses than the clearest statements about their economic and social benefits"[14] (Zuiderzeevereniging 1916: 53). The Zuiderzee Society organized a public exhibition on the 1916 floods, demonstrating how the Zuiderzee Works could help to prevent such disasters in the future. In a national newspaper, Vissering (1916: 5) argued that the floods taught the Dutch a hard and painful lesson and expressed his hope that the people and their representatives would now finally be willing to seriously consider Zuiderzee plans. Smeenge, in one of his parliamentary speeches, noted that adopting the Zuiderzee bill would lead people in other countries to admit that the Dutch are "a small people, that dares to take risks in fearful times" (Zuiderzeevereniging 1920: 334). In doing so, they convinced the larger public as well as politicians of the benefits of the project. The bill was ratified on March 21, 1918, in the Lower House and on June 13 in the Senate, with which the Zuiderzee Act was adopted.

---

[13]Original Dutch quotation: "Maar men maakt van eene provincie toch geen dambord!".

[14]Original Dutch quotation: "Al is het Zuiderzeegevaar niet de eenige en zeker niet de voornaamste beweegreden voor de afsluiting en gedeeltelijke drooglegging van de Zuiderzee, het spreekt krachtiger tot de menigte dan de helderste betoogen omtrent economische en sociale voordeelen.".

For the implementation of the Zuiderzee Works, Lely established a new organization (Dienst der Zuiderzeewerken) instead of handing this task to Rijkswaterstaat. Lely also created a Zuiderzee Council that would oversee the implementation of the project, in which Lely himself took seat as its president and Vissering and H. Colijn were appointed as its vice presidents. Other members of the Zuiderzee Society, among whom were Smeenge and Beekman, took seats in this council as well. The implementation was not without hurdles. The economic recession that followed the war put an extra burden on the advocates of the Zuiderzee Works to re-legitimize the expenditures in these circumstances. When agricultural engineer and former director of the Heidemaatschappij H.J. Lovink (1866–1938) published the report of his state commission that was asked to reevaluate the costs and benefits of the Zuiderzee Works, this came at a convenient time for Colijn, who had just been installed as Minister of Finance and was faced with huge budget cuts on the Zuiderzee Works. Supported by Lovink's conclusions that the Works would require about 380 million but that the benefits would be larger than anticipated (Lovink 1924), Colijn drafted a bill that accelerated the implementation of the Zuiderzee Works by allowing the government to take out a loan, which was adopted in 1926.

2.3.2 Resistance from the Zuiderzee Fisheries

The Zuiderzee Works would have a huge impact on the fishing industry. The issue was, however, not as contentious as technical-financial concerns and initially did not stir up much unrest in the policymaking process on the Zuiderzee Works.

In his fifth technical note, Lely mentioned that the construction of the Afsluitdijk would terminate the Zuiderzee fishing industry. For the members of the Zuiderzee Society, this sacrificing of the Zuiderzee fishing industry was outweighed by the benefits created by the Zuiderzee Works. As Lely argued in his notes, "the fishing grounds will be replaced by new land with rich clay soil, of which the yearly profits will surpass that off the fishing industry many times"[15] (Zuiderzeevereniging 1892: 21). The 1892 state commission was, however, of the opinion that the issue had not been appropriately addressed in Lely's notes. In its recommendations, the committee therefore included different proposals to help Zuiderzee fishermen cope with the negative impacts of the Zuiderzee Works (Staatscommissie 1894). The commission suggested life-long retirement pensions for fishermen above the age of 55. Younger fishermen should be supported in relocating their activities to the North Sea; they should be provided with new boats and exempted from the obligation to pay North Sea port fees. The costs of these measures were estimated at 4.5 million guilders.

Minister Lely, in his 1901 bill, accepted the state commission's proposals and created a budget of 4.5 million for compensation measures. Even so, Lely's bill stirred

[15]Original Dutch quotation: "dan wordt het vischgebied vervangen door eene nieuwe provincie vruchtbare kleigronden, waarvan de jaarlijksche bruto-opbrengst vele malen die der visscherij zal overtreffen".

some unrest in fishing communities. A schoolteacher in Volendam, B. Demmer, set up a "General Committee of Zuiderzee fisheries" to conduct a counter-research based on data collected from the fisheries industry itself (Zuiderzeevereniging 1905a). The Demmer commission concluded that instead of the proposed 4.5 million, 14 million would be more appropriate to compensate the losses of the industry. This collective initiative was, however, an exemption. Further collective efforts were troubled by an old conflict that deeply divided Zuiderzee fishermen.

As early as 1500, fishermen from the affluent western province of North-Holland, forced by salinization of their western fishing spots, expanded their sailing grounds to the eastern part of the Zuiderzee. Their superior towed-fyke trawlers contrasted sharply with the traditional practices of eastern fishermen, who worked with off-shore fishing nets in fishing spots allocated by their guilds (Dorleijn 1982).[16] Western fishermen neglected this system and freely sailed across the allocated eastern fishing spots (Ypma 1962). When the Amsterdam-Rijnkanaal—which directly connected Amsterdam to the North Sea—was constructed and North Sea competition increased, western fishermen retreated to the Zuiderzee and the east-west conflict was reinforced. The two sides became organized into different lobby groups, with eastern groups advocating a ban on towed fyke nets to prevent over-exploitation of fish stocks while western groups tried to prevent such a ban (Bossaers 1987).

This strife shaped fishermen's reactions to Zuiderzee plans. For example, when C. Redeke (1907), expert advisor on the committee for sea-fisheries (College voor Zeevisscherijen) who was asked to look at the issue of compensation to fishermen, published his report in which he concluded that the Zuiderzee fishing industry was not in decay, his findings were rebutted by eastern fishermen who experienced declining catches as a result of an increase in the use of fyke-trawlers. Another example is the in 1911-established Zuiderzee Fisheries Council that represented different local fishing associations. This council was criticized heavily both by western and by eastern fishermen for over-representing the other side. While Demmer had successfully organized a collective reaction with his inventory and report in 1901, 12 years later, when he called for a collective protest after the 1913 queen's speech addressed the Zuiderzee Works, no reaction followed. Even Lely's 1918 bill did not give rise to joint efforts on behalf of fishermen.

The Zuiderzee Society did not take a clear stand on the issue of compensation. On the one hand, they believed Zuiderzee fishermen would benefit from the Zuiderzee Works, as new job opportunities would open up in construction activities and later on the reclaimed lands. On the other hand, they felt responsible for helping fishermen make the transfer. In reaction to the 1901 Demmer report, the association commissioned two studies into the matter. The Neeb-Committee established in 1905 sketched a gloomy picture; not only did Zuiderzee fishermen live in deprived conditions, but the small-scale industry was destined to be swallowed by up-scaling processes anyway (Zuiderzeevereniging 1905a). Rather than providing them with

[16]Fyke nets are triangle-shaped fishing nets. They can be "fixed" in between poles in fishing waters, like the eastern fishermen used to do, to be hauled in every couple of days. Western fishermen used lightweight fyke nets that they attached to their ships while sailing.

financial compensation, the commission advised to create "new circular routes to make sure the places where other businesses are being conducted are easily accessible and to open up opportunities to accumulate knowledge"[17] (idem: 246). Only for the elder generation and for widows, it argued that compensation was justified. In 1906, the Society published the results of another study, undertaken by the agricultural development company "Heidemaatschappij",[18] which concluded that a freshwater fish stock could emerge in the IJssel Lake if the waters were managed properly. Based on both studies, the Zuiderzee Society argued that "one should admit that closing off and reclaiming parts of the Zuiderzee will not terminate a thriving business; on the contrary, the fishing grounds could be exploited in a new and more rational manner"[19] (Zuiderzeevereniging 1906: 6). After 1906, the association paid little attention to the issue of fishermen anymore.

Lely's 1918 bill only included the general promise that 4.5 million guilders would be reserved for compensation to fishermen through measures that would be devised by an expert commission. In the parliamentary debates, the fisheries issue surfaced only a couple of times. The bill was criticized for being vague with respect to who would receive compensation and how much (Zuiderzeevereniging 1920: 432–437), but a rather clumsy debate followed on what the preferences and needs of fishermen actually were; do they prefer compensation over new job opportunities, or do they want to continue fishing? The only decision taken on the issue was to require parliament to approve of future compensation rules in order to keep a check on their implementation.

After the adoption of Lely's bill, fishermen started to have trouble attracting loans for new materials, as their suppliers were not sure anymore the fishermen could repay their debts under uncertain future circumstances (Ritter 1932). In reaction to complaints about this problem, the expert commission on compensation rules set up by Lely, which was headed by Vissering, arranged to erect a credit agency to come to the aid of fishermen.

This expert commission also worked on a bill for compensation to fishermen. Vissering found that there was no legal right for compensation, as the Zuiderzee was not private property and therefore fishermen were formally not expropriated (Ter Veen 1935). For reasons of compassion, he proposed to supplement fishermen's wages up to their mean income level over 1915–1917 and to grant elderly fishermen

[17]Orgininal Dutch quotation: "De commissie ziet veel meer in het creëren van nieuwe verbindingswegen om gemakkelijk de plaatsen te kunnen bereiken waar andere bedrijven worden uitgeoefend, en de opening van de gelegenheid om de kennis te vermeerderen" (246).

[18]This engineering and development company focused on cultivating land in harsh environments. The company changed its name into "Acradis" in 1997 and has developed into an internationally known advisory and construction company in civil, but still mainly hydraulic, engineering. The company was, for example, involved in the levee reconstruction activities after hurricane Katrina in New Orleans. .

[19]Orginal Dutch quotation: "Wanneer men dit alles nagaat, zal men moeten toegeven dat het argument, als zoude door de afsluiting en gedeeltelijke drooglegging der Zuiderzee de visscherij als een bloeiende tak van bedrijf te gronde worden gericht, onhoudbaar is; integendeel zou juist die tak van visscherij op eene geheel nieuwe en dan meer rationele wijze kunnen worden uitgeoefend.".

a pension. Other measures included a right to free education for fishermen and their children, preferential hiring, loans to start new businesses, and licenses to fish on the new IJssel Lake. When the bill was sent to parliament in 1923, the first signs of the economic depression of the 1930s just started to be felt. The incumbent Minister of Water Management restricted the duration of financial allowances to three years for people in between the ages of 18 and 25 and to five years for people between 25 and 35 to reduce the budget (Kamerstukken II 1924/1925: 30). Parliament adopted this Zuiderzee Support Act in 1925.

Fishermen were hugely disappointed about the amount of financial support they received through this bill. They felt let down by Lely, who had promised them compensation for their losses while in reality it turned out to be very difficult to apply for income supplements. Likewise, the pensions for the elderly were based on the Poverty Act and were therefore very low (Bossaers 1987; Dorleijn 1985). It was in reaction to this collective disillusionment of the Zuiderzee fisheries industry that schoolteacher Demmer's 1901 General Committee was revived. In 1928, it organized a protest meeting in which over 600 fishermen from eastern as well as western provinces participated. Again, the issue of compensation was discussed in parliament (Kamerstukken II 1927/1928). In 1930, Demmer organized yet another demonstration in which over 1400 people participated, including fishermen from all over the Zuiderzee region, representatives of fisheries-related companies, members of local governments, and even some members of parliament. Demmer's arguments with respect to the expectations created by Lely and the Zuiderzee Society were given widespread attention in the newspapers (cf. Algemeen Handelsblad 1930). These concerted efforts led to an amendment of the Zuiderzee Support Act, adopted by a large majority in parliament (Kamerstukken II 1930/1931: 7). The executive agency dealing with compensation was replaced by a new state agency to establish more fair and transparent procedures, which ultimately improved support for fishermen.

2.4 The Implementation of the Zuiderzee Works

It took the Zuiderzee Society a long time to gain support for their Zuiderzee plans in politics. Its implementation proved another sweeping procedure. The process was not only held back by technical disputes and social concerns described in this chapter, but the economic crisis in the 1930s, financial shortages in the post-Second World War period, and ecological protests in the 1970s had further delayed the reclamation of the polders. In fact, the fourth Zuiderzee polder was so often postponed that in 2003, the Dutch government decided to abandon the plan to empolder this area altogether (VROM 2004). Despite all these struggles, the Afsluitdijk stood the test of time; it protected the northern part of the country against the devastating storm surge of 1953 that caused a flooding disaster in the southern part of the country, and over the years the Dutch have benefitted greatly from the added value created by the cultivation of their new self-made province (Van der Geest et al. 2008: 27). Some may therefore argue that the policymaking process took much longer than needed, as forward-looking experts were held back by "partisan" conflicts of interests.

At the same time, this chapter also demonstrates that the experts involved in the Zuiderzee Society were very influential and that they not only determined the technical, but to an important extent also the social and distributive aspects of the Zuiderzee Act that was adopted in 1918. In hindsight, project expenditures far exceeded the costs calculated by these experts (Thijsse 1972). Instead of the 190 million guilders budgeted in Lely's technical notes, recent estimates are that the Dutch government has spent a total amount of three billion guilders on the project (Van der Geest et al. 2008: 26). Also, the fishermen's issue was not adequately addressed by the Zuiderzee Society—which did its best to portray the Zuiderzee fisheries as an industry in decay—and it took subsequent parliamentary actions to correct this problem. Based on these insights, it could be argued that expert-influence reached too far into the "political" domain of decision-making in the case of the Afsluitdijk.

Taking into consideration that the Zuiderzee case allows for these multiple interpretations of the role of experts in the policymaking process, it first and foremost calls for a careful analysis of expert-influence. What factors constituted the influence of experts in this process, and what did this imply for the way in which the distributive aspects of the Zuiderzee Works policy were dealt with?

2.5 Conclusion

This chapter aimed to describe the role of experts in the formation of the safety approach in Dutch flood governance, examplified by the implementation of the Zuiderzee Works, in order to then analyze what this expert-influence implied for the extent to which distributive aspects of safety measures were recognized in the political decision-making process.

The reconstruction of the policymaking process underlying the Zuiderzee Works in this chapter on the one hand demonstrates the importance of the self-organizing capacity of experts. Organized into the Zuiderzee Society, the experts involved in the policymaking process were part of a rising Dutch elite that actively endeavored for a greater role of experts in public affairs. Nearly all of them were either bankers with a profound interest in water management or civil engineers with political ambitions. Most of them had served in the colonial administration. What knitted them together was an ideology in which democracy and meritocracy collided. They favored a leading role of scientists in public policymaking, and many were in favor of general suffrage. They therefore believed that the projects experts proposed needed support, not only from policymakers but also from the public at large. To generate this approval, experts carefully outlined not only the costs involved with their projects but more importantly also the (economic and social) benefits that would be created by their plans. It was through these efforts that the Zuiderzee Society managed to generate public and political support for the construction of the Afsluitdijk.

On the other hand, this chapter described how the political context of the Netherlands at the beginning of the 20th century also provided the experts of the Zuiderzee Society with the room to develop their expertise in the political and public domain. Economic and social deprivation undermined the liberal state doctrine at the end of the 19th century and created receptivity for the ideas of this new elite who presented an alternative to dominant laissez-faire principles. Most members of the Zuiderzee Society belonged to dissident religious denominations. Mennonites (Doopsgezinden) and Remonstrants were in the majority. As religious outsiders, they did not fall prey to traditional religious divides.

Based on these insights, it is concluded that the combination of expert organization and contextual circumstances gave the Zuiderzee Society the nearly unchallenged status of "expert-group" in the policymaking process on the Zuiderzee Works. The politically "embedded" character of expert-influence meant that strong interactions were created between experts and policymakers in Dutch flood governance. As a result of these interactions, the lines between "politics" and "science" blurred, which was most vividly demonstrated by Lely's threefold occupancy of a ministerial post in Dutch government. Because of the close relationship between the Zuiderzee Society and policymakers, these experts gained structural access to the policymaking process.

The interaction between the experts of the Zuiderzee Society and policymakers produced a clear understanding of the problem of floods, in which floods were constructed as an external risk that posed a threat to a socially and economically vibrant Dutch society. The central state was appointed as the appropriate actor to deal with this threat. In the social-technocratic ideology of the experts involved in the Zuiderzee Society, dealing with the local-level impacts of a plan that so clearly served the public interest was seen as a "political" issue that belonged to the domain of ad hoc democratic decision-making. Thus, through their expert status in Dutch flood governance; the Zuiderzee Society strongly influenced not only the technical but also the social and distributive aspects of the Zuiderzee Works policy.

Considering these strong bonds between policymakers and experts, and the coherent policy discourse that was developed through their interaction, it may be expected that alternative viewpoints and interests were blocked from the policymaking process. However, this chapter shows otherwise. It demonstrates that the problem definition offered by the Zuiderzee Society could be challenged in the democratic policymaking and decision-making process. As soon as the Zuiderzee Society published Lely's technical notes, civil engineers at Rijkswaterstaat, for example, openly disagreed with the financial and technical assumptions underlying the calculation of national costs and benefits balance in Lely's notes. Bearing in mind that the protagonists of the Afsluitdijk strongly resented the inclusion of partisan interest in public governance, it is even more striking that claims on behalf of Groningen about increased flood risks for this northern province and concerns about Zuiderzee fishermen about losing their jobs also emerged in the policymaking process.

The counter positions that were formulated in the policymaking process of the Zuiderzee Works were also taken on board in the policymaking process. Arguments that challenged the technical or financial aspects of the Zuiderzee Works were usually dealt with by the experts themselves by commencing further research and adjusting the plan in accordance to new findings. Based on the concerns voiced by Rijkswaterstaat engineers, Lely adjusted the technical assumptions behind his plans. He also pledged additional research into the effects of the construction of the Afsluitdijk for flood risks in Groningen. And when the Zuiderzee fisheries industry managed to overcome its internal divisions and started to organize their occupational interest more collectively—a development that cannot be disconnected from the extensions of general suffrage in that period—a commission was installed to see to the needs of Zuiderzee fishermen.

References

Algemeen Handelsblad (1930) De Zuiderzeesteunwet. *Algemeen Handelsblad* 103(33639): Available via the National Library of the Netherlands. http://resources3.kb.nl/010660000/pdf/DDD_010661634.pdf. Accessed 15 Aug 2013

Baneke D (2011) Synthetic Technocracy: Dutch Scientific Intellectuals in Science, Society and Culture, 1880–1950. Br J Hist Sci 44(160):89–113

Beck U (1992) Risk society—towards a new modernity. Sage publications, London

Beyen M (2008) Een gezond oorlogskind. Parlementaire discussies over de afsluiting en de drooglegging van de Zuiderzee, 1918. In: T Sintobin (ed) *Getemd maar rusteloos. De Zuiderzee verbeeld – Een multidisciplinair onderzoek* (pp 73–89). Verloren, Hilversim

Bossaers KWJM (1987) *Zuiderzeevissers – Geschiedenis van de Belangenorganisaties van de vissers op de Zuiderzee en IJsselmeer*. De Walburg Pers & Stichting voor het Bevolkingsonderzoek in de Drooggelegde Zuiderzeepolder, Zutphen

Braat WC (1932) De archeologie van de Wieringenmeer – een bijdrage tot de geschiedenis van het ontstaan van de Zuiderzee. N.V. Boekhandel en Drukkerij, voorheen E.J. Brill, Leiden

Cleintuar GL (1982) Wisselend Getij – Geschiedenis van de Zuiderzeevereniging 1886-1949. De Walburg Pers, Zutphen

Cleintuar GL (1990) Cornelis Lely. In: GHL Tiesinga (ed) *Het Zuiderzeeproject, voor- en tegenstanders, plannenmakers en uitvoerders*, pp 23–26. De Walburg Pers & Stichting voor het Bevolkingsonderzoek in de drooggelegde Zuiderzeepolders, Lelystad

Danner HS (1992) Droogmakerijen in de zeventiende eeuw. In: Beenakker JJJM, Danner HS (eds) *Strijd tegen het water – Het beheer van land en water in het Zuiderzeegebied*, pp 45–61. Walburg Pers & Zuiderzeemuseum Enkhuizen, Zutphen

Den Hoed P, Keizer A (2007) *Op steenworp afstand - op de brug tussen wetenschap en politiek*. Amsterdam University Press, Amsterdam

Dorleijn P (1982) Van gaand en staand want. De zeilvisserij voor en na de afsluiting van de Zuiderzee, deel II. Stichting voor het Bevolkingsonderzoek in de drooggelegde Zuiderzeepolders. Van Kampen & Zn, Amsterdam & Bussum

Dorleijn P (1985) Van gaand tot stand want. De zeilvisserij voor en na de afsluiting van de Zuiderzee, deel IV. Stichting voor het Bevolkingsonderzoek in de drooggelegde Zuiderzeepolders. Van Kampen & Zn, Amsterdam & Bussum

Drentje J (2011) Thorbecke Revisited - De rol van het doctrinaire liberalisme in de Nederlandse politiek. BMGN - BMGN - Low Countries Historical Review 126(4):3–28

Fennema M, Rhijnsburger J (2007) Dr. Hans Max Hirschfeld - Man van het grote geld. Uitgeverij Bert Bakker, Amsterdam

Fuchs JM, Simons WJ (1972) De Afsluitdijk: Recht Door Zee. Knoop & Niemeijer, Haren

Habermas J (1996, translated by W. Regh) Between facts and norms—contributions to a discourse theory of law and democracy. The Massachusetts Institute of Technology Press, Cambridge, MA

Jansma K (1954) Lely – Bedwinger der Zuiderzee. H.J. Paris, Amsterdam

Kamerstukken II 1917/1918a, 67e vergadering, nr. 63, Handelingen, pp 1878–1897
Kamerstukken II 1917/1918b, 69e vergadering, nr. 63, Handelingen, pp 1958–1977
Kamerstukken II 1924/1925, 281, nr. 5
Kamerstukken II 1927/1928, 266, 72e vergadering, Handelingen, pp 2082–2106
Kamerstukken II 1930/1931, 35, nr. 1
Krips-Van der Laan H (1999) Woord en daad – De zoektocht van Derks Roelfs Mansholt naar een betere samenleving. Van Gorkum & comp. B.V, Assen
Lintsen, H. (1980). Ingenieurs in Nederland in de Negentiende Eeuw – Een Streven naar Erkenning en Macht.'s-Gravenhage: Martinus Nijhoff
Lovink HJ (1924) Verslag der Commissie tot het instellen van een hernieuwd onderzoek naar de baten, welke van de afsluiting en droogmaking der Zuiderzee mogen worden verwacht. Algemeene landsdrukkerij, s-Gravenhage
Molhuysen PC, Blok PJ (1918) Tak van Poortvliet. Nieuw Nederlandsch biografisch woordenboek, Deel 4. A.W. Sijthoff, Leiden. Available via DBNL. http://www.dbnl.org/tekst/molh003nieu04_01/molh003nieu04_01_1898.php. Accessed 22 Feb 2013
NIDI (2003) Bevolkingsatlas van Nederland – Demografische ontwikkelingen van 1850 tot heden. Nederlands Interdisciplinair Demografisch Instituut & Uitgeverij Elmar B.V, Rijswijk
Quack HPG (1915) Herinneringen uit de levensjaren van Mr. H.P.G. Quack 1834-1913, 2nd edn. P.N. van Kampen & Zoon, Amsterdam
Rayner S (2012) Uncomfortable knowledge: the social construction of ignorance in science and environmental policy discourses. Economy and Society 41(1):107–125
Redeke C (1907) Rapport over onderzoekingen betreffende de visscherij in de Zuiderzee, ingesteld in de jaren 1905 en 1906. Ministerie van Landbouw, Nijverheid en Handel & Gebroeders van Cleef, 's-Gravenhage
Ritter CW (1932) De economische betekenis van de Zuiderzee. In: Colijn H (ed) De Zuiderzee, een herinneringswerk. Scheltema's & Holkema's Boekhandels en Uitgevers, Amsterdam, pp 39–50
Rosanvallon P (2008, translated by A. Goldhammer). Counter-democracy—politics in an age of distrust. Cambridge University Press, Cambridge
Staatscommissie (1894) Verslag der Staatscommissie benoemd bij Koninklijk Besluit van 8 September 1892, no. 21, tot het instellen van een onderzoek omtrent eene afsluiting en eene droogmaking van de Zuiderzee. 's-Gravenhage: Gebroeders Belinfante
Ter Laan K (1949) Multatuli en twee van zijn discipelen Mansholt en De Raaf. E.J. Brill, Leiden
Ter Veen HN (1925) De Haarlemmermeer als kolonisatiegebied. P Noordhoff, Groningen
Ter Veen HN (1935) Zuiderzee en staat. De Economist 84(1):727–739
Thijsse JTh (1972) Een halve eeuw Zuiderzeewerken, 1920–1970. H.D. Tjeenk Willink, Groningen
Toussaint B (2009) Training the rivers and exploring the coasts. Knowledge evolution in the Netherlands in two engineering fields between 1800 and 1940. Phys Chem Earth 34(3):132–141
Van Blom D (1917) Van zee tot meer en land. De Gids 81:124–162
Van der Brink A, Molema M (2008) The origins of Dutch rural planning: a study of the early history of land consolidation in the Netherlands. Plann Perspect 23(4):427–453
Van der Geest L, Berkhof M, Meijer M (2008) Het hoofd boven water – Tweehonderd jaar investeren in waterwerken. NYFER. Available via Delta Commissie. http://www.deltacommissie.com/doc/Nyfer%20Het%20Hoofd%20Boven%20Water%20.pdf. Accessed 21 Mar 2013
Van der Ham W (2007) Verover mij dat land. Lely en de Zuiderzeewerken. Amsterdam, Boom
Van der Houven-van Oordt HC, Vissering G (1901) De economische beteekenis van de afsluiting en drooglegging der Zuiderzee. E.J. Brill, Leiden
Van der Houven-van Oordt HC, Vissering G (1998) De economische beteekenis van de afsluiting en drooglegging der Zuiderzee. E.J. Brill, Leiden
Van der Vleuten E (2004) Infrastructures and societal change, a view from the large technical systems field. Technol Anal Strateg Manag 16(3):395–414
Van Urk A, Prince Claus of The Netherlands, Oudshoorn H, Schultz B, Zijderveld P (1999) Sustainable Development of Delta's, Sustainable Development of Delta's Conference 1998, Proceedings. Delft university Press, Delft

Vissering G (1916) Een Harde Les. Algemeen Handelsblad, 16 Jan 1916: 5. Available via the National Library of the Netherlands. http://kranten.kb.nl/view/paper/id/ddd%3A010652535%3Ampeg21%3Ap005%3Aa0005. Accessed 9 Dec 2013

Vogel KM (2008) 'Iraqi Winnebagos™ of death': imagined and realized futures of US bioweapons threat assessments. Sci Publ Policy 35(8):561–573

Ypma YN (1962) Geschiedenis van de Zuiderzeevisserij. Report no 27. Publicaties van de Stichting voor het bevolkingsonderzoek in de drooggelegde Zuiderzeepolders, Amsterdam

Zuiderzeevereniging (1892) Resultaten van Technisch Onderzoek 1886–1892. Boekhandel en Drukkerij, voorheen E.J. Brill, Leiden

Zuiderzeevereniging (1905a) Verzameling rapporten uitgegeven door de Zuiderzeevereniging, deel I – De Zuiderzeevisscherij. Rapport eener commissie tot onderzoek. Boekhandel en Drukkerij, voorheen E.J. Brill, Leiden

Zuiderzeevereniging (1905b) Verzameling rapporten van de Zuiderzeevereniging, deel II. Rapporten aan Z.E. den Minister van Waterstaat, Handel en Nijverheid met Nota van Beantwoording der Zuiderzeevereniging. Boekhandel en Drukkerij, voorheen E.J. Brill, Leiden

Zuiderzeevereniging (1906) Verzameling van rapporten uitgegeven door de Zuiderzeevereniging, deel III. Rapport van de Nederlandsche Heide-Maatschappij over de Zoetwatervisscherij in het toekomstige IJsselmeer en in de wateren der droog te leggen polders. Boekhandel en Drukkerij, voorheen E.J. Brill, Leiden

Zuiderzeevereniging (1916) De Watervloed van 13–14 januari 1916. Boekhandel en Drukkerij, voorheen E.J. Brill, Leiden

Zuiderzeevereniging (1920) Handelingen en bijlagen van de beide kamers der Staten-Generaal betreffende het ontwerp van wet tot afsluiting en droogmaking van de Zuiderzee. Boekhandel en Drukkerij, voorheen E.J. Brill, Leiden

Chapter 3
Engineering Space: Spatial Flood Risk Management in the Netherlands

3.1 Introduction

Internationally, spatial measures are gaining popularity as part of a more "integrated" or "risk-based" approach to flood governance. In integrated flood risk management, the aim is no longer only to reduce the chance of flooding, as is done in a traditional "safety" approach to floods, but explicitly also to minimize the impacts of floods (Klijn et al. 2008; Bubeck et al. 2012; Hegger et al. 2014). The intention is to find an optimal balance between "hard" protection and "soft" spatial measures.

Incorporating spatial measures in flood risk management can have major implications for the traditional distribution of costs and responsibilities in flood governance. Existing studies have pointed to transfers in governance responsibilities from the central to the regional and local level as a result of the implementation of spatial measures (e.g., Johnson and Priest 2008; Meijerink and Dicke 2008; Rijke et al. 2012). They also found changing allocations of costs and benefits in flood governance under a spatial approach to floods (e.g., Merz et al. 2010; Butler and Pidgeon 2011; Penning-Rowsell and Pardoe 2012; Paudel et al. 2015).

Up to now, little attention has been paid to the role of knowledge in the transition from a safety approach to integrated flood risk management. While most analyses underscore the fact that integrated flood risk management requires new forms of knowledge to support integrated organizational arrangements (e.g., Macdonnell 2008; Birkmann and Von Teichman 2010; Herk et al. 2011; Vink et al. 2013), a detailed account of what new knowledge requirements actually emerge in this process has not yet been made. This chapter focuses on knowledge requirements for distributive decision-making in particular. Does the evaluation of distributive implications of spatial policies require other forms of expertise than standardly incorporated under

Parts of this chapter have been published in: Bergsma, E. (2016) Changed knowledge requirements for spatial flood governance. *Ecology and Society*, 21(4): 40, https://doi.org/10.5751/ES-08952-210440.

E. Bergsma, *From Flood Safety to Spatial Management*, Water Governance - Concepts, Methods, and Practice, https://doi.org/10.1007/978-3-319-96716-5_3

safety institutions? And how can these new forms of knowledge be organized? These questions are analysed in a case study on the Netherlands.

This chapter builds on an institutional perspective. In line with the broader outset of this book, the shift to spatial measures in Dutch flood governance is seen as a process of gradual institutional adaptation, in which existing safety institutions do not suddenly "turn" but gradually move towards a spatial approach, in response to day-to-day changes in processes of political agency and collective sense-making (Dewulf et al. 2004, 2007). Conceptually, this shift is grasp in terms of the Policy Arrangements Framework (PAF), which distinguishes between different dimensions of institutions that can change: a discourse, an actor, a resources and a rules dimension (Van Tatenhove et al. 2000; Arts and Van Tatenhove 2004; Arts et al. 2006). This chapter focuses on the role knowledge in this process of institutional change. It investigates what type of knowledge supported the shift to spatial measures in Dutch flood governance and what this implied for the evaluation of the distributive impacts of spatial measures in the decision-making process.

For the analysis, the policymaking processes underlying three key spatial policies that have been discussed in Dutch flood governance since the 1990s have been reconstructed. The first is the Room for the River program, which started in the late 1990s and aims to create more space for (flood)water in the Dutch national landscape. Second, attempts to set up an insurance scheme for flood damage will be discussed, as they reflect the intention to place more emphasis on flood-resilient spatial planning at the local level. The third is the adoption of the concept of "multi-layered safety" in the Second Delta Program, which was implemented in 2008 and highlights the importance of flood-proof spatial planning in the Netherlands. The reconstructions are based on an analysis of parliamentary records, and additional scientific and policy documents. The results have been checked and fine-tuned in meetings with key government organizations in Dutch flood governance.

The chapter is structured as follows. Section 3.2 briefly sketches the evolution of the safety approach in Dutch flood governance over the course of the 20th century. Section 3.3 provides the analyses of the three selected policy measures. For each of these policies, the policymaking process is reconstructed based on an analysis of parliamentary records and additional documentation that provided insight into the role of expert-knowledge in these policy developments. Section 3.4 reflects on the findings and draws conclusions.

3.2 The Evolution of Dutch Flood Governance in the 20th Century

Being at the basis of the Dutch safety approach, engineers continued to play an important role in the evolution of Dutch flood governance over the course of the 20th century. One of the major actors in the field was Rijkswaterstaat. Established in the late 18th century under French influence as part of the military apparatus responsible for centralizing Dutch water management, Rijkswaterstaat transformed from a military to a civil-engineering organization in the 20th century and gradually rose as the expert body on Dutch water management (Lintsen 1980). Under the leadership of Rijkswaterstaat's engineers, the safety approach further institutionalized in Dutch flood governance (Van den Brink 2009).

3.2.1 Standardization of the Safety Approach in the 1950s–1960s

An important factor in the development of the safety approach in Dutch flood governance was the disastrous flood of 1953. That year, a major storm surge breached the levees protecting the south-eastern delta of the country. More than 200,000 hectares of land flooded, over 1800 people were killed, and the total damage loss equaled 5.2 billion euro (Dijke 2013: 215). This event reinforced the focus of Dutch flood governance on technical flood protection.

After the 1953 flood, Rijkswaterstaat was handed the responsibility to develop and implement a new flood protection system for the Netherlands. Rijkswaterstaat set up a "Delta Committee" to work on a "Delta Plan" for Dutch flood protection. Most importantly, this Delta Plan encompassed the embankment of estuaries in the flooded south-eastern areas (also known as the "Delta Works"). But their plan went further and also included measures to raise and strengthen other parts of the coastal levee system. As a guiding principle, the Delta Committee adopted so-called "exceedance norms", which specified the maximum height of storm surge elevations that coastal levees should be able to stand. These norms were based on an evaluation of the expected costs involved with a levee failure. For the most economically vital areas of the Netherlands, the norm of 1/10,000 was adopted, meaning that levees should be able to ward off storm surge elevations with a statistical chance of occurring once every 10,000 years. This meant that coastal levees should provide protection against water level elevations of up to five meters above average. Taking into account that levees would not immediately fail when such high water levels occurred, the actual protection these norms provided was thought to be higher. For 1/10,000 norm areas, the protection standard was, for example, estimated at 1/125,000 (denoting a flood change of no more than once every 125,000 years). For other coastal levees, norms and safety standards were set lower, depending on the expected damage that would be prevented by levee reinforcement.

With this strong national-planning tradition in Dutch flood protection, emergency management and damage compensation also largely remained a national state responsibility. At first, governmental damage compensation mainly targeted the direct impacts of flood defenses. As the previous chapter demonstrated, compensation was offered to the Zuiderzee fishermen for the losses they suffered as a result of the construction of the Afsluitdijk. The 1958 Delta Act, which implemented the Delta Works, also arranged for damage compensation to groups who were disadvantaged by the construction of the Delta Works. For example, in article 5, section 4d, of this law, damage compensation was arranged for communities that had to make way for levee construction, while article 7 dealt with compensation for value reductions of property and article 8 set up a specific arrangement for the fisheries industry (Stb. 1958, 246).

Flood damage compensation was differently arranged. Before 1953, flood damage could be insured in the private market. After the 1953 flood, insurers decided to stop covering flood damage, as the risk had become too great. Their decision to leave the flood insurance market created a void in responsibilities for flood damage compensation. In practice, this void was filled by central-governmental institutions. Over the years, a generous damage compensation practice emerged in which the national government not only compensated for the negative effects of flood protection works, but increasingly also for the damage done by flood events themselves (Kuks 2004: 96). Every time a major flood struck, different administrative arrangements were set up that offered damage compensation to specific groups. For example, the Ministry of Agriculture arranged damage compensation for farmers, the ministry of Economic Affairs compensated the damage of private businesses, and individuals were usually compensated with the money collected through public fundraising actions and/or directly from the national treasury by the Ministry of Interior.

After the construction of the Zuiderzee Works, the safety approach standardized in Dutch flood governance. Under the safety institutions that developed, Dutch flood governance evolved as a technically planned and heavily regulated governmental responsibility focused on providing safety to floods (Gupta et al. 2016). With more than 50% of the national surface below sea level and about 70% of all properties located in these areas (Van der Brugge et al. 2005: 164–176), the Dutch are highly dependent on technical protection for safety against floods, and for damage compensation if this protection fails. The costs of flood protection are high. Dike reinforcements, the upkeep of secondary levee system and the daily management of the water system require huge efforts. These activities are funded through public taxes, and costs are roughly split between the central government and regional water boards who raise separate taxes (MIM 2014: 3). The increased role of the national government in private flood damage compensation only adds costs to the national-level.

In the Netherlands, the high investments that the safety approach requires have always been justified based on long-term cost-benefit analyses in which the expected loss of life and amount of (material and immaterial) damage resulting from floods was traded-off against the costs of investment in flood protection. These cost-benefit analyses are normally produced by independent state committees that bring together the necessary technical-engineering, economic and state planning expertise to support reliable calculations. The 1953 Delta Committee, for example, consisted of 12 civil-engineers, one agronomist and the famous Dutch national economist Jan Tinbergen.

3.2.2 The Incorporation of Ecological Expertise in the 1970s–1980s

In the 1970s, in line with the growth of (global) environmentalism in this decade, concerns also arose over the ecological impact of the Dutch flood protection policy. To provide flood safety, the Dutch had rigorously changed the natural landscape, distorting hydrological and ecological processes and causing damage to natural and cultural heritage.

From this concern, protest was raised against the safety norms proposed by Delta Committee in the 1970s. In response to these protests, a new committee was set up to develop safety norms for riverine areas, taking into account the impacts of these norms on the environment. This Becht committee included Rijkswaterstaat deputies, civil and hydraulic engineers employed at regional water management authorities and provincial water management units, land-use planners and environmental representatives (Commissie Rivierdijken 1977). Using the probability of increased river runoff as the basis for developing riverine safety norms, and taking into account the possible ecological damage of river levee reinforcement, this committee advised to lower the Delta Committee norms in some river regions to a standard of 1/1.250 (Heezik 2007: 220–221). These norms were adopted in subsequent legislation.

In the 1980s, the first "spatial" component was added to the Dutch portfolio of flood governance strategies. In 1986, ecologists and landscape architects presented a spatial vision (called "*Plan Ooievaar*") on the riverine area in the Netherlands in which rivers' natural floodplains would be restored and used for environmental restoration (De Bruin et al. 1987). While this plan called to increase space for water, it was not seen as an alternative to structural flood protection; rather, it served to counteract the negative environmental impacts of flood control projects on the environment (Van Leussen and Meijerink 2014).

Since the 1970s, ecological expertise has been included as an additional source of knowledge in Dutch flood governance (Van den Brink 2009). With this so-called "ecological turn", Dutch flood governance has not only become known for its ground-breaking engineering technologies, but also for its "polder-model" of decision-making in which different (economic, safety, and ecological) stakes have been made part of the process (Disco 2002). Spatial measures were first forwarded under this ecological turn. However, it was only in the 1990s that spatial measures were for the first time discussed as an alternative to flood defense (Vis et al. 2003).

3.3 The Implementation of Spatial Measures in the Netherlands

In the 1990s, high water levels in some of the country's main rivers accumulated into riverine floods in 1993 and a near-flooding disaster 1995. In addition, recurring rainfall extremes between 1998 and 2002 led to a number of flood incidents that caused repeated damages. In response to these incidents, the national government had to dig deep into the national budget for flood damage compensation. At the same time, expectations about increased future flood risks under climate change put pressure on the national government to invest in levee reinforcement. Considerations about spatial measures arose from this specific context (Van Buuren et al. 2012; Jong and Van den Brink 2013). In this section, the policymaking processes underlying three spatial policy proposals are analyzed in more detail.

3.3.1 Room for the River

In the 1990s, the Dutch were caught by surprise by a series of (near) flooding incidents. In 1993, the Meuse River burst through its banks in the southern province of Limburg, flooding one-fifth of the provincial surface. The floods put more than 700,000 properties underwater, of which many were built in unembanked areas in the river's winter bed. Similar circumstances occurred in 1995. While a large flooding disaster was prevented, its anticipation led to one of the largest evacuation projects ever undertaken in Dutch history. In the late 1990s and early 2000s, a series of rainfall extremes caused multiple instances of flooding throughout the country, which resulted in a substantial amount of crop and property damage.

The immediate reaction to the 1993 flood event was a "typical" Dutch one. In parliament, the event was referred to as an "environmental disaster" that called for "national solidarity" with its victims (Kamerstukken II 1993/1994a: 5). In 1995, a Delta Act Large Rivers was enacted to speed up dike reinforcements in riverine areas. Especially after the 1995 floods, however, this standard governance response was also criticized. Rijkswaterstaat and regional water management authorities were increasingly faced with the difficult task of fighting off floods for a society that continued to allow spatial developments in flood-prone areas. While throughout history Dutch engineers had always called for building stronger levees to keep the water out, they now started arguing that more space should be reserved for floodwater to better deal with the impacts of climate change. This vision was, for example, present in an influential report of the Advisory Committee on Water Management in the 21st Century, an engineer-led committee appointed to explore the future challenges of Dutch water management. In their influential report, this committee argued that the Dutch needed to start accepting floods from time to time and focus on reducing their consequences by reserving more space for water in the national landscape (Commissie WB21 2000).

The view that flood protection had reached its limits in a new context of climate change started to prevail in parliament as well. Dike reinforcement was criticized for only building up the value at risk to flooding, and the prolongation of this "unsustainable" practice was seen as uneconomical (Kamerstukken II 1996/1997a: 3). In 1996, a Room for the River Policy Guideline was adopted that, for the first time, diverted Dutch flood governance away from its protection doctrine (VROM and VenW 1996). The guideline restricted building activities in the winter bed of some of the country's major rivers. It arranged that economic activities were only allowed if they were riverine-tied (e.g., shipping) or represented a major national-economic interest, and it required constructions in the winter bed to meet a certain safety standard to minimize flood damage. To reimburse foregone revenues lost because of these Room for the River planning restrictions, a special damage compensation arrangement was erected (Kamerstukken II 1996/1997b).

Over time, and in absence of large-scale flood events, the planning restrictions imposed by the Room for the River Policy Guideline were experienced as too burdensome. Under the leadership of a conservative-liberal government in 2006, Dutch parliament amended the 1996 guideline with a new Large Rivers Policy Guideline. This new guideline permitted more developments in the winter bed by letting go of safety requirements for damage mitigation (VenW and VROM 2006). It also emphasized the individual responsibility of project developers and users of winter bed areas to deal with the impacts of flood events. The damage compensation arrangement erected in 1996 was dismantled, and instead, the 2006 guideline stated that development initiators and residents would themselves be held responsible for taking sufficient precautionary measures to mitigate flood damage and to shoulder the costs of flood damage in the event of a flood (idem.: 7). Flood insurance was proposed as a means to assist citizens and businesses in acting on this new responsibility (Kamerstukken II 2005/2006a).

In 2007, Room for the River projects were incorporated under a structural planning decision that gave river-widening measures a more definite status in Dutch flood governance. The national government also strengthened its competences in this field. The new Spatial Planning Act adopted that year granted the national government a right to specify the ground conditions for and overrule municipal spatial plans to achieve national water goals (Needham 2005).

The development of the Room for the River policy described in this section reveals a gradual change in the Dutch policy discourse on floods (see also Wiering and Arts 2006; Immink 2007). In this new policy discourse, the understanding of floods as a natural hazard was replaced by a focus on floods as an environmental risk partly driven by unwise location choices. This implied that floods could no longer be tackled through collective state protection and that measures were needed to address the anthropogenic causes underlying the risk. The spatial measures that were implemented shifted the attention from reducing the chance of flooding to reducing the impacts of flood events. These changes were instigated by "institutionalized" experts in Dutch flood governance: engineers and macro-economists. These experts were also prominently involved in the development of new spatial measures, whose national cost-benefit analyses were now also used to identify economically efficient land-use options to create more space for water in the Netherlands.

Consequently, the organizational structure in Dutch flood governance did not change much at first. Floods were still dealt with through top-down (river and landscape) engineering solutions, for which key responsibilities were assigned to the national government, which designated and financed Room for the River projects, specified the building conditions in project areas, and compensated the damage of those negatively affected by flood control projects. These changes in the discourse on floods did, however, create openings for the implementation of new costs and responsibility structures in flood governance. When a more conservative-liberal cabinet rose to power, a decentralization of responsibilities for flood protection and damage compensation was legitimized through this discourse. While flood insurance was initially seen as a means to help these actors cope with their new responsibilities in flood governance, the following section demonstrates the difficulties encountered in this area.

3.3.2 Flood Damage Compensation

In contrast to many other (European) countries, flood insurance is not generally available in the Netherlands. Instead, the Dutch national government often covered private losses after a major flood event. It often used different administrative arrangements to compensate the damage of different actor-groups affected by a flood (Duin and Mesu 1995). For example, the Ministry of Agriculture arranged damage compensation for farmers, the Ministry of Economic Affairs compensated the damage of private businesses, and individual flood losses were usually reimbursed by the Ministry of Interior. This was generally accepted as a reasonable and fair way to tailor governmental damage compensation to the specific needs of affected groups (Kamerstukken II 1993/1994b, 1994/1995a). However, the recurrent character of flooding in the 1990s and 2000s challenged this practice. On the one hand, the arbitrary character of the arrangements was called into question, with different levels of damage compensation being offered to different groups (Kamerstukken II 1995/1996). On the other hand, this practice was seen to assert a moral pressure on the Dutch government to pay out every time a flood struck, an obligation that was no longer deemed tenable under the increased flood risks posed by climate change, not only because it would become unaffordable, but also because it reduced incentives to mitigate flood damage at the local level (Kamerstukken 1994/1995b).

In policy discussions on damage compensation after the 1993 and 1995 floods, flood insurance has often been proposed as a means to improve the existing damage compensation practice (Kamerstukken II 1994/1995c, 1994/1995d). Flood insurance would not only provide a more systematic method of damage compensation, as it operates under fixed rules, but it would also incentivize the implementation of loss reduction measures at the local level because people could lower their insurance premiums by taking these measures. However, flood insurance has always encountered much resistance in the Netherlands. When in 1995 a bill was drafted that would set up a public-private flood insurance scheme, this bill was sharply criticized by the Council of State for conflicting with the national government's constitutional duty of care for sustaining the quality of the Dutch living environment (Kamerstukken II 1995/1996). Among policymakers, the general feeling also was that flood damage compensation involved questions of national solidarity that should not be left to the market but should be settled in parliament instead (idem.).

With insurance being a no-go area, Dutch policymakers in 1998 adopted a "Disasters and Serious Accidents (Compensation) Act". The law intended to balance the goals of legal security, national solidarity, and individual responsibility (Kamerstukken II 1996/1997c). Legal security was provided by specifying the conditions under which the law could be activated by the Minister of Interior, which was in the case of a large-scale riverine flooding or a flood event with a similar magnitude. National solidarity was ensured by preserving the minister's freedom to determine the specific compensation rules on a case-by-case basis. Individual responsibility was addressed by only arranging for partial compensation of flood damage and by excluding losses that were reasonably insurable or could be attributed to careless actions.

Applications of this law have been much contested. While the law has been activated for different kinds of events, it has not always been activated under recurrent circumstances because of the structural character and therefore predictability of the problem that actors can anticipate. In addition, compensation rules have been amended on a case-by-case basis (Kamerstukken II 1998/1999, 2000/2001). Because of this, the law has been criticized for failing to provide financial security as well as incentives for damage mitigation (Faure and Hartlief 2001; Botzen and Van den Bergh 2008). In 2004, an advisory committee concluded that the division of responsibilities for flood damage compensation in the Netherlands remained unclear under the new compensation act (Commissie Tegemoetkomingen bij Rampen en Calamiteiten 2004). In an official reaction to this committee's report, the Dutch government emphasized that citizens are primarily responsible for shouldering their own damage in the event of a flood and that the national government only has a legitimate role in cases of large-scale, socially disruptive flooding disasters (Kamerstukken II 2005/2006b: 1). Since that time, opportunities to implement an insurance scheme for flood damage have again been explored (Kamerstukken II 2005/2006b, 2013). In 2003, an agrarian rainfall insurance scheme was successfully implemented and several attempts have been undertaken to launch a general insurance scheme for private flood losses (Kok et al. 2014).

Under the traditional damage compensation regime in the Netherlands, policy decisions were evaluated based on a national-level consideration of harms and injuries suffered from a flood event. The national government assumed a large part of the costs for damage compensation. This damage compensation practice fit in well with the policy arrangements of a safety approach to floods, which already assigned a key role to the national government in flood protection. However, as responsibilities for protection shifted to the local level under Room for the River projects, this traditional flood damage compensation practice lost its natural appropriateness.

While a complete conversion to an insurance system was always resisted in the Netherlands because of conflicts with underlying normative beliefs and traditions in Dutch flood governance, elements of flood insurance have gradually been incorporated in the rules-dimension of the institutional arrangement in Dutch flood governance. They were discussed as a logical consequence of the spatial planning approach implemented through Room for the River projects, which continued to be justified based on its national cost-benefit balance. Because of this, the implications of shifts in responsibilities produced by implementing flood insurance were not specified in the policymaking process and did not generate much attention in the decision-making process. In parliamentary discussions on flood insurance between 1998 and 2014, the question of what individuals could actually do to reduce their exposure to flood risks was raised only once. This question was answered by a simple reference to the option to "upscale" buildings, but this left aside a whole bunch of questions relating to how individuals would do this, whether this was even allowed in the strictly regulated spatial planning structure of the Netherlands, and who would pay for such measures.

Strikingly, little new expertise was brought into the policy discussions on flood insurance, although especially in the last two decades, knowledge on the relationship between flood insurance and spatial measures quickly developed in the Netherlands. Scholars connected to geography and spatial economy departments of Dutch knowledge institutions have, in particular, analyzed and discussed the prospects and drawbacks of using insurance as a policy instrument under a risk-based approach to flood risk management in the Netherlands (e.g., Vrijling et al. 2008; Botzen et al. 2010; Aerts and Botzen 2011; Seifert et al. 2013; Kok et al. 2014; De Moel et al. 2014; Jongman et al. 2014; Paudel et al. 2015). While the expertise is available, this knowledge has not found its way into the policymaking process.[1] Because of this, the distributive consequences of emphasizing local responsibilities for dealing with the impacts of floods largely remained unspecified in the policy discussions on flood damage compensation.

3.3.3 The Second Delta Program

In 2007, a Second Delta Committee was appointed to analyze the state of the Dutch flood protection system. Seating, amongst others, an economist, a civil engineer, a climate expert, a landscape architect, and the director of a large dredging company, this committee concluded that the Dutch flood protection system not only failed to meet its current standards, but that these standards were too low to adequately prepare the water system for the impacts of climate change in general. When the committee published its findings in 2008, the report functioned as an alarm bell in Dutch flood governance (Verduin et al. 2012).

In response to the findings of this committee, a Second Delta Program was erected in 2008, which has since then been updated every year. Through this program, the Dutch government outlines its water goals and specifies its policy strategies. A principle that guides this program is that of "multi-layered safety". Developed by water management professionals, this principle distinguishes between three safety layers to function as the pillars of policymaking on floods: a first layer of flood prevention, a second layer of climate-proof spatial planning, and a third layer of emergency management (Meijerink and Dicke 2008). With the Second Delta Program, the Dutch government formulated so-called "delta decisions" that together embody this principle of multi-layered safety.

[1] Discussion meeting at the Dutch Ministry of Infrastructure and the Environment, October 12, 2015, The Hague.

The first delta decision concentrates on water safety, which is seen as the primary pillar of Dutch flood governance (VenW et al. 2009). With this decision, the Dutch government calls for a revaluation of the costs and benefits involved with national flood protection. The performance of this revaluation was delegated to a group led by economists at the engineering agency Deltares, an organization where many former Rijkswaterstaat employees have been employed since Rijkswaterstaat underwent several reorganizations that aimed to transform it from an engineering agency to a more diverse group of specialist working on water management.[2] The Deltares group developed a new method to calculate what they termed "economically efficient flood protection standards" for the Netherlands—defined as the point at which further investments in dike reinforcement exceed the gained benefits of mitigated flood damage (Eijgenraam et al. 2014). Justified in reference to their cost-efficiency (Kamerstukken II 2010/2011, 2011/2012, 2013/2014, 2015/2016), this new norm-setting method was adopted by parliament in 2015. The new norms that will be calculated through this method will provide an equal basic level of protection to each Dutch citizen, expressed as an annual chance of being killed by a flood of no more than 1/100,000. This standard is matched by no other country in the world. In addition, safety norms will be set higher in certain (mainly riverine) areas while they are lowered for other areas. The logic behind this is an economical one; according to the Second Delta Program, safety standards will be set higher in areas where there are a lot of people or where there is a lot of value at risk, because the benefits of flood protection outweigh the costs of strengthening flood protection (Delta Program 2015: 16).

Another delta decision focuses on spatial adaptation. With this decision, the Dutch government outlines spatial measures as a means to address the second (climate-proof spatial planning) and third (emergency management) safety layers. While spatial measures are described as a sensible policy strategy in general, their importance is specifically underscored for areas where the cost of reinforcing flood protection are unreasonably high compared to the value protected. In these areas, the Delta Program calls for so-called "smart combinations" in which spatial measures are implemented instead of dike reinforcement (Kamerstukken II 2015/2016). Because spatial planning is locally organized in the Netherlands, spatial measures require "shared ownership" and "self-regulation" (Delta Program 2014: 8). The search for "smart combinations" in low-risk areas means that the Dutch government places more responsibility for taking precautionary measures against floods and for mitigating flood damage on the shoulders of individuals and businesses in low-risk areas than in high-risk areas, as high-risk areas continue to be protected through collectively funded protection works. However, this policy implication was not recognized at all in the policy discussions on the Delta Program. In reaction to an assessment of the Organization for Economic Co-operation and Development (OECD 2014), which concluded that Dutch citizens were not well aware of and therefore ill-prepared for the flood risks they face, there has been some discussion on the role of citizens in Dutch flood governance. However, these discussions concentrated on the need to inform

[2]Discussion meeting at Rijkswaterstaat, February 15, 2016, Lelystad..

citizens about the emergency relief measures they can take during a flood rather than on their role in preventing flood damage through spatial measures (Kamerstukken II 2013/2014, 2015/2016).

With the Second Delta Program, elements of an integrated flood risk management approach have been incorporated under a single policy framework. These are generally accepted as innovative elements in the Dutch polder-model for flood governance that will help prepare the country for the future impacts of climate change. While as part of this approach the basis of expertise of important water governance authorities such as Rijkswaterstaat and regional water boards has been diversified, the policy choices underlying this program continue to depend on institutionalized engineering expertise, which is now "outsourced" to organizations like Deltares where many former Rijkswaterstaat engineers have been employed. The choice to protect an area with protection or spatial measures is based on the national cost-benefit analysis performed by the Deltares group. Under this national-level focus, however, the local-level implications of these policy choices have not always been clearly recognized in policy discussions.

3.4 Conclusion

This chapter analyzed the shift from safety to spatial measures in Dutch flood governance. Understood as a process of gradual institutional adaptation, the aim was to study the role of experts in the evolution and implementation of spatial measures in the Netherlands, and the implications of this expert-involvement for the recognition and discussion of the distributive implications of spatial measures. To this end, the policymaking processes underlying three spatial policy proposals were reconstructed and analyzed. Based on the analyses, three conclusions are drawn.

First, the analysis shows that the shift to spatial measures in Dutch flood governance was a long-term process. The first consideration of spatial measures emerged in response to environmental concerns about flood control in the 1970s and 1980s, which gradually opened up the safety discourse. However, it was only in the 1990s that spatial measures were for the first time discussed a flood control strategy in itself. Spatial measures were proposed as an alternative to structural flood protection, which could be more cost-effective in in a future of increased costs of flooding caused by climate change.

Arguments about the relative effectiveness of spatial measures were not brought in by a new type of expert. The impacts of climate change were highlighted by institutionalized engineering experts who, under the safety approach, were trusted with the responsibility of protecting the Netherlands against flooding but who felt this responsibility was increasingly at jeopardy by the unwarranted development of flood-prone areas and climate change. The new spatial planning policies they devised aimed to tackle these problems (e.g., the Room for the River Program), but they did so in a way that largely followed a "safety logic"; spatial measures were justified through economic cost-benefit analyses and were implemented top-down. However, by highlighting the importance of adjusting spatial planning to new flood risks, a new policy discourse gradually developed in which human settlement choices were identified as an additional cause underlying flood risks. This new discourse opened up possibilities for change at the other dimensions of the policy arrangement. Notably, with the identification of anthropogenic causes, new rules that emphasized local-level responsibilities for flood protection and damage compensation could be justified (see also Meijerink and Dicke 2008; Bergsma et al. 2012; Van Buuren et al. 2012).

Second, the analyses in this chapter point out that new knowledge requirements emerged in this process of institutional adaptation. The previous chapter explained how under safety institutions, flood governance strategies were justified based on a national consideration of their costs and benefits. This chapter demonstrated that in order to evaluate the distributive implications of spatial measures, knowledge is needed that provides insight in the local-level distributive implications of spatial measures because costs and responsibilities shift to local and individual actors under a safety approach. For example, what measures are actually available for local actors to make flood-resilient location choices or to flood-proof their buildings in the second, spatial planning layer of Dutch flood protection set out in the Second Delta Plan, and what costs are associated with such measures? And what do the principles of "shared ownership" and "self-regulation" in this spatial planning layer mean for the right to damage compensation, particularly for people living in low- to medium risk areas where spatial planning solutions are thought to be most efficient?

Third, this chapter has shown that knowledge on the local-level distributive implications of spatial measures was not part of the policymaking process. While insights about the local-level policy implications of spatial measures have been developed within geography and economy departments of Dutch universities, this knowledge did not find its way into the policymaking process. A major cause behind this can be linked to the actor dimension of the policy arrangement in Dutch flood governance, where the institutionalized relationship between engineers and policymakers blocked the incorporation of new types of experts and new types of expertise in the policymaking process.

References

Aerts JCJH, Botzen WJW (2011) Climate change impacts on pricing long-term flood insurance: a comprehensive study for the Netherlands. Glob Environ Change 21(3):1045–1060

Arts B, Leroy P, Van Tatenhove J (2006) Political modernisation and policy arrangements: a framework for understanding environmental policy change. Publ Organ Rev 6(2):93–106

Arts B, Van Tatenhove J (2004) Policy and power: a conceptual framework between the 'old' and 'new' policy idioms. Policy Sci 37(3/4):339–356

Bergsma E, Gupta J, Jong P (2012) Does individual responsibility increase the adaptive capacity of society? The case of local water management in the Netherlands. J Resour Conserv Recycl 64:13–22

Birkmann J, Von Teichman K (2010) Integrating disaster risk reduction and climate change adaptation: key challenges—scales, knowledge, and norms. Sustain Sci 5(2):171–184

Botzen WJW, van den Bergh JCJM (2008) Insurance against climate change and flooding in the netherlands: present, future, and comparison with other countries. Risk Anal 28(2):413–426

Botzen WJW, Van den Bergh JCJM, Bouwer LM (2010) Climate change and increased risk for the insurance sector: a global perspective and an assessment for the Netherlands. Nat Hazards 52(3):577–598

Bubeck P, Botzen WJW, Aerts JCJH (2012) A review of risk perceptions and other factors that influence flood mitigation behavior. Risk Anal 32(9):1481–1495

Butler C, Pidgeon N (2011) From "flood defence" to "flood risk management": exploring governance, responsibility, and blame. Environ Plann C: Gov Policy 29(3):533–547

Commissie Rivierdijken (1977) Rapport Commissie Rivierdijken. Commissie Rivierdijken, Den Haag

Commissie Tegemoetkomingen bij Rampen en Calamiteiten (2004) Solidariteit met beleid. Aanbevelingen over financiële tegemoetkomingen bij rampen en calamiteiten. Available via Officiele Bekendmakingen. https://zoek.officielebekendmakingen.nl/kst-27157-63-b1.pdf. Accessed 8 Sep 2015

Commissie WB21 (2000) Water beleid voor de 21e eeuw—Geef water de ruimte en de aandacht die het verdient. Commissie Waterbeheer 21e eeuw: Den Haag. Available via TU Delft. http://repository.tudelft.nl/view/hydro/uuid%3A102e013a-1357-4087-b9f3-387f877c793f/. Accessed 9 Sep 2015

De Bruin D, Van Hamhuis D, Nieuwenhuijze L (1987) Ooievaar: de toekomst van het rivierengebied. Stichting Gelderse Milieufederatie, Arnhem

De Moel H, Van Vliet M, Aerts JCJH (2014) Evaluating the effect of flood damage-reducing measures: a case study of the unembanked area of Rotterdam, the Netherlands. Reg Environ Change 14(3):895–908

Delta Program (2014) Synthesedocument Ruimtelijke adaptatie Achtergronddocument B3. Available via Delta Commissaris. http://www.deltacommissaris.nl/deltaprogramma/documenten/publicaties/2014/09/16/deltaprogramma-2015-achtergronddocument-b3. Accessed 15 Sep 2015

Delta Program (2015) Werk aan de Delta. De beslissingen om Nederland veilig en leefbaar te houden. Available via Delta Commissaris. http://www.deltacommissaris.nl/deltaprogramma/documenten/publicaties/2014/09/16/deltaprogramma-2015. Accessed 15 Sep 2015

Dewulf A, Craps M, Dercon G (2004) How issues get framed and reframed when different communities meet: a multi-level analysis of a collaborative soil conservation initiative in the ecuadorian andes. J Community Appl Soc Psychol 14(3):177–192

Dewulf A., François G, Pahl-Wostl C, Taillieu T (2007) A framing approach to cross-disciplinary research collaboration: experiences from a large-scale research project on adaptive water management. Ecology & Society 12(2):14. http://www.ecologyandsociety.org/vol12/iss2/art14/ (accessed June 24, 2015).

Dijke SC (2013) De overstromingsverzekering in Nederland. Het Verzekerings-Archief 4:214–224

Disco C (2002) Remaking "Nature" the ecological turn in Dutch water management. Sci Technol Human Values 27(2):206–235

Duin MJ, Mesu JJ (1995) Wie zal dat betalen? Overheid, schadevergoeding en aansprakelijkheid bij wateroverlast. Bestuurskunde 4(8):361–369

Eijgenraam C, Kind J, Bak C, Brekelmans R, Den Hertog D, Duits M, Roos K, Vermeer P, Kuijken W (2014) Economically efficient standards to Protect the Netherlands against flooding. Interfaces 44(1):7–21

Faure M, Hartlief T (2001) Schade als gevolg van natuurverschijnselen: de betekenis van de WTS. Milieu & Recht 28(10):240–246

Gupta J, Bergsma E, Termeer CJAM, Biesbroek GR, Van den Brink M, Jong P, Klostermann JEM, Meijerink S, Nooteboom S (2016) The adaptive capacity of institutions in the spatial planning, water, agriculture and nature sectors in the Netherlands. Mitig Adapt Strat Glob Change 21(6):883–903

Heezik AAS (2007) Strijd om de rivieren – 200 jaar rivierenbeleid in Nederland. Dissertation, Technische Universiteit Delft, The Netherlands

Hegger DLT, Driessen PPJ, Dieperink C, Wiering M, Raadgever GTT, Van Rijkswick HFMW (2014) Assessing stability and dynamics in flood risk governance: an emperically grounded research approach. Water Resour Manage 28(12):4127–4142

Immink I (2007) Established and recent policy arrangements for river management in The Netherlands: an analysis of discourses. Dissertation, Wageningen University, The Netherlands

Johnson CL, Priest SJ (2008) Flood risk management in England: a changing landscape of risk responsibility? Int J Water Resour Dev 24(4):513–525

Jong P, Van den Brink M (2013) Between tradition and innovation: developing Flood Risk Management Plans in the Netherlands. J Flood Risk Manage. https://doi.org/10.1111/jfr3.12070

Jongman B, Koks EE, Husby TG, Ward PJ (2014) Increasing flood exposure in the Netherlands: implications for risk financing. Hydrol Earth Syst Sci 14:1245–1255

Kamerstukken II (1993/1994a) 23568, nr. 3. Sdu Uitgevers: Den Haag

Kamerstukken II (1993/1994b) 23564, nr. 10. Sdu Uitgevers: Den Haag

Kamerstukken II (1994/1995a) 24071, 5. Sdu Uitgevers: Den Haag

Kamerstukken II (1994/1995b) 24071, nr. 8. Sdu Uitgevers: Den Haag

Kamerstukken II (1994/1995c) 24071, 21, herdruk. Sdu Uitgevers: Den Haag

Kamerstukken II (1994/1995d) 24071, nr. 27. Sdu Uitgevers: Den Haag

Kamerstukken II 1995/1996, 24640, nr. 1. Sdu Uitgevers: Den Haag

Kamerstukken II (1996/1997a) 18106, nr. 71. Sdu Uitgevers: Den Haag

Kamerstukken II (1996/1997b) 18106, nr. 81. Sdu Uitgevers: Den Haag

Kamerstukken II (1996/1997c) 25159, nr. 5. Sdu Uitgevers: Den Haag

Kamerstukken II (1998/1999) 24071, nr. 36. Sdu Uitgevers: Den Haag

Kamerstukken II (2000/2001) 24071, nr. 56. Sdu Uitgevers: Den Haag

Kamerstukken II (2005/2006a) 30462, nr. 4. Sdu Uitgevers: Den Haag

Kamerstukken II (2005/2006b) 29668, nr. 11. Sdu Uitgevers: Den Haag

Kamerstukken II (2010/2011) 31710, nr. 20. Sdu Uitgevers: Den Haag

Kamerstukken II (2011/2012) 31710, nr. 26. Sdu Uitgevers: Den Haag

Kamerstukken II (2013/2014) 27625, nr. 328. Sdu Uitgevers: Den Haag

Kamerstukken II (2015/2016) 34300-J, nr. 6. Sdu Uitgevers: Den Haag

Klijn F, Samuels P, Van Os A (2008) Towards flood risk management in the EU: state of affairs with examples from various European countries. Int J River Basin Manag 6(4):307–321

Kok M, Van Vliet L, Klijn F (2014) Deltafact - Verzekeren van overstromingsschade. Available via Delta Proof. http://deltaproof.stowa.nl/pdf/Verzekeren_van_overstromingsschade?rId=54. Accessed 14 Sep 2014

Kuks SMM (2004) Water governance and institutional change. Dissertation, University of Twente, The Netherlands

Lintsen H (1980) Ingenieurs in Nederland in de Negentiende Eeuw – Een Streven naar Erkenning en Macht. 's-Gravenhage: Martinus Nijhoff

Macdonnell RA (2008) Challenges for integrated water resources management: how do we provide the knowledge to support truly integrated thinking? Int J Water Resour Dev 24(1):131–143

Meijerink S, Dicke W (2008) Shifts in the public-private divide in flood management. Water Resour Dev 24(4):499–512

Merz B, Hall J, Disse M, Schumann A (2010) Fluvial flood risk management in a changing world. Nat Hazards Earth Syst Sci 10(3):509–527

MIM (2014) Letter of the Dutch Ministry of infrastructure and the environment on water security, 2 June 2014, reference no. IENM/BSK-2014/96343. Ministerie van Infrastructuur en Milieu, Den Haag. Available via Hoogwaterbeschermingsprogramma. http://www.hoogwaterbeschermingsprogramma.nl/documenten+openbaar/kennis+en+innovatie/handlerdownloadfiles.ashx?idnv=448454. Accessed 3 Jun 2018

Needham B (2005) The New Dutch spatial planning act. Plan Pract Res 20(3):327–340

OECD (2014) Water governance in the Netherlands: fit for the future? Organisation for Economic Co-operation and Development, Paris. Available via Rijksoverheid. https://www.rijksoverheid.nl/documenten/rapporten/2014/03/17/oecd-studies-on-water-water-governance-in-the-netherlands-fit-for-the-future. Accessed 15 Sep 2015

Paudel Y, Botzen WJW, Aerts JCJH, Dijkstra TK (2015) Risk allocation in a public-private catastrophe insurance system: an actuarial analysis of deductibles, stop-loss, and premiums. J Flood Risk Manage 8(2):116–134

Penning-Rowsell EC, Pardoe J (2012) Who benefits and who loses from flood risk reduction? Environ Plan C: Gov Policy 30(3):448–466

Rijke J, Van Herk S, Zevenbergen C, Ashley R (2012) Room for the river: delivering integrated river basin management in the Netherlands. Int J River Basin Manage 10(4):369–382

Seifert I, Botzen WJW, Kreibich H, Aerts JCJH (2013) Influence of flood risk characteristics on flood insurance demand: a comparison between Germany and the Netherlands. Nat Hazards Earth Syst Sci 13(7):1691–1705

Van Buuren A, Klijn E, Edelenbos J (2012) Democratic legitimacy of new forms of water management in the Netherlands. Int J Water Resour Dev 28(4):629–645

Van den Brink M (2009) Rijkswaterstaat on the horns of a dilemma. Eburon, Delft

Van der Brugge R, Rotmans J, Loorbach D (2005) The transition in Dutch water management. Reg Environ Change 5(4):164–176

Van Leussen W, Meijerink S (2014) Environmentalism, renewed concern for flood safety, and the Europeanization of Dutch water policies, 1970–2010. In: Lonnquest J, Toussaint B, Manous J Jr, Ertsen M (eds) Two centuries of experience in water resources management—A Dutch-U.S. retrospective. Institute for Water Resources, U.S. Army Corps of Engineers & Den Haag: Rijkswaterstaat & Ministry of Infrastructure and the Environment, Alexandria, pp 247–286

Van Tatenhove J, Arts B, Leroy P (eds) (2000) Political modernisation and the environment: the renewal of environmental policy arrangements. Kluwer Academic Publishers, Dordrecht & Boston & London, pp 53–69

VenW, VROM (2006) Beleidslijn Grote Rivieren. Available via Rijkswaterheid. https://www.rijksoverheid.nl/documenten/kamerstukken/2006/07/26/beleidslijn-grote-rivieren. Accessed 4 Feb 2016

VenW, VROM, LNV (2009) Beleidsnota Waterveiligheid 2009–2015. Available via Rijkswaterheid. https://webcache.googleusercontent.com/search?q=cache:P24mWzsNBoIJ:https://www.rijksoverheid.nl/binaries/rijksoverheid/documenten/kamerstukken/2009/12/22/b-beleidsbota-waterveiligheid-pdf/b-beleidsbota-waterveiligheid.pdf+&cd=1&hl=nl&ct=clnk&gl=nl. Accessed 4 Feb 2016

Verduin SH, Meijerink SV, Leroy P (2012) How the second delta committee set the agenda for climate adaptation policy: a Dutch case study on framing strategies for policy change. Water Altern 5(2):469–484

Vink MJ, Boezeman D, Dewulf A, Termeer CJAM (2013) Changing climate, changing frames. Dutch water policy frame developments in the context of a rise and fall of attention to climate change. Environ Sci Policy 30:90–101

Vis M, Klijn F, De Bruin KM, Van Buuren A (2003) Resilience strategies for flood risk management in the Netherlands. Int J River Basin Manage 1(1):33–40

Vrijling JK, Eijgenraam CJJ, Kok M (2008) Verzekeren tegen grote overstromingen. Expertise Netwerk Waterveiligheid, Utrecht

VROM, VenW (1996) Beleidslijn Ruimte voor de Rivier. Available via the Dutch Ministry of Infrastructure and the Environment. http://publicaties.minienm.nl/documenten/beleidslijn-ruimte-voor-de-rivier. Accessed 3 Oct 2015

Wiering MA, Arts BJM (2006) Discursive shifts in Dutch river management: 'Deep' institutional change or adaptation strategy? Hydrobiologia 565(1):327–338

Chapter 4
From Levees to Flood Insurance: The Spatial Turn in US Flood Governance

4.1 Introduction

This chapter investigates the relationship between experts and political actors in 20th century US flood governance, during which a transition was made from a "safety" to a new "spatial" approach. The aim is to analyze the role of experts in this transition, as well as to analyze the effects of their influence on the distributive decision-making (dealing with the allocation of costs and responsibilities) underlying it. What experts were involved in this transition, what was their influence on the development of the US policy discourse on floods, and how have governance arrangements in US flood governance, of which the distribution of costs and responsibilities is part, changed as a result?

In the US, a spatial approach to floods emerged in the 1940s and was officially implemented in the 1960s. In order to fully grasp this policy transition, this chapter analyses its emergence against the background of the safety approach that characterized US flood governance in the first decades of the 20th century, and studies the institutionalization of the US spatial approach to floods in the second half of the 20th century. The analysis focuses on several moments in the policymaking process during which key distributive decisions were made. For each of these moments, the main policy developments are mapped out based on an examination of (scientific and policy) literature on US flood governance. Transcripts of Congressional debates and committee hearings were used to understand how expert-knowledge fed into the policymaking process and how it shaped the understanding and consideration of the distributive choices underlying spatial measures of political actors in the democratic decision-making process.

Parts of this chapter have been published in: Bergsma, E. (2016) Geographers Versus Managers: Expert Influence on the Construction of Values Underlying Flood Insurance in the United States. *Environmental Values*, 25(6): 687–705.

E. Bergsma, *From Flood Safety to Spatial Management*, Water Governance - Concepts, Methods, and Practice, https://doi.org/10.1007/978-3-319-96716-5_4

The chapter is structured as follows. Section 4.2 briefly sketches the emergence and partial institutionalization of the safety approach in US flood governance in the early decades of the 20th century. Section 4.3 describes how, from the 1930s onward, this safety approach was challenged and changed by a new "spatial planning" discourse on floods, which was formalized into policy in the late 1960s. Section 4.4 covers the gradual institutionalization of this approach in US flood governance over time, up to the start of the 21st century. Section 4.5 discusses the role of experts in the transition to spatial measures in US flood governance. The conclusion reflects on the influence of these experts on distributive decision-making underlying this turn.

4.2 The Engineers' Era (1900–1920)

Just like in the Netherlands (see Chaps. 2 and 3), the first centralization tendencies in US flood governance emerged at the start of the 20th century (Wright 2000). Up to the 1900s, it was left to local communities and states to fund flood protection and bear the costs of flood damage. Motivated by overseas successes in the construction of the Panama Canal—a Saint-Simonian inspired project that was started by the French in 1881 but taken over by the US Army Corps of Engineers in 1904, under whose management the canal was finished below budget and two years ahead of schedule—the Army Corps engineers started to develop plans to control domestic rivers as well. Most of these plans targeted the Mississippi delta, an important economic growth region.

At first, the plans of the Army Corps stayed confined to improving the Mississippi's function as a navigational route for commercial shipping because at that time, the federal government was only allowed to invest in "internal" state affairs if these investments benefited the national economy (Layton 1986). Over the years, Army Corps plans stretched out to other goals and targets, including flood protection. The growing role of the Army Corps in water management threatened the position of independent civil engineers who used to provide their services to US state governments and the federal government. In an effort to reground the status of their expertise, these civil engineers started to organize themselves into associations through which they developed engineering plans for the public good. This gave a great boost to the engineering profession. The number of engineers in the US rose from 7000 in 1880 to 226,000 in 1930 (Barry 1997: 290). Involved in an open competition for federal funding, civil and Army Corps engineers now openly criticized each other's plans (see Barry 1997 for a vibrant description of this rivalry). In this competition, engineers increasingly emphasized flood protection as an additional benefit of their plans (Tarlock 2012).

In 1917, the federal government for the first time allocated funds specifically to flood control. Limited in scope to the Mississippi and Sacramento Rivers, the 1917 Flood Control Act arranged that for every one dollar spent on flood protection by "local interests," the federal government would pay at least two (Arnold 1988: 14). After devastating Mississippi floods in 1927, a new Flood Control Act was adopted in 1928 that authorized one of the largest flood protection projects ever undertaken in the lower Mississippi valley. This time, no local contributions were required at all (Barry 1997).

Under the leadership of engineers, the federal government became increasingly involved in flood protection in the first decades of the 20th century. This growing federal role in local state affairs met with resistance. It clashed with dominant political values, like local autonomy and individual responsibility. However, there was no available alternative to controlling floods at that time.

4.3 The Emergence of a Spatial Planning Approach

4.3.1 The Geographer's Floodplain Management Approach (1930–1950)

Federal investments in flood control were reinforced under Franklin D. Roosevelt's New Deal Administration, which launched a series of state projects to tackle rising unemployment levels during the Great Depression in the 1930s (Wright 2000). Water management proved conductive to this interventionist approach; flood protection became unaffordable for local communities during the economic recession, and federally-funded flood control projects created much-needed jobs (Macdonald et al. 2012). In 1936, over $310 million was appropriated to subsidize 250 different flood control projects throughout the nation (Wright 2000: 11).

Roosevelt's interventionist regime was a key topic of concern for social scientists connected to the Chicago school of behavioral sciences. Set up by professor Charles E. Merriam as a Social Sciences Research Council at the University of Chicago in 1929, this school worked with a rational-actor perspective to explore the opportunities for adjusting "negative" and promoting "rational" or "wise" behavioral patterns through governmental incentives. According to Chicago School members, rational incentives often provided a cost-efficient alternative to direct governmental interventions (Simon 1985). Within the Chicago school, a group of geographers developed a particular interest in natural resource management. The increased role of the federal government in flood control, which was accompanied by growing federal expenditures on disaster relief, was a mounting concern, especially for these geographers. With a Congress willing to fund flood control projects to create employment, these geographers considered it their mission to develop a method to better weigh the costs of flood protection against its benefits. Led by professor Harlan H. Barrows of the university's geography department, this group accentuated spatial planning interventions as a more cost-efficient way to reducing flood risks (Macdonald et al. 2012).

Members of the Chicago school developed close ties with Roosevelt's administration, which took a great interest in natural resource management as a means to combine multiple policy goals (e.g., job creation, resource stability, and environmental improvement, cf. Reuss 1992). Both Merriam and Barrows held a position in the Roosevelt administration's National Planning Board that was established within the department of Public Works (Hinshaw 2006: 15). When Barrows joined the Water Resources Committee that was formed within this planning board, his work provided a notable exception to the engineering focus displayed by this committee. In a 1936 report, engineers on this committee called for the construction of more levees, river outlets, and reservoirs to prevent floods. Barrows, however, called for a spatial planning approach to floods through which people would be guided away from high-risk areas to minimize flood impacts. He argued that "[i]f it would cost more to build reservoir storage than to prevent floodplain encroachment, all relevant factors considered, the latter procedure would appear to be the better solution" (Barrows 1936, cited in Reuss 1992: 116). This chapter laid out the beginning of the US spatial approach to flood governance.

Barrows's ideas were further elaborated by his student, Gilbert F. White. In his PhD thesis titled "The Human Adjustment to Floods", White (1945) argued that levees and other "hard" protection measures could in principle be an effective way to reduce flood losses, but not in all areas. Particularly in medium- to low-risk areas, the costs of investing in flood protection did not outweigh the benefits gained by protection. However, as White argued, such rational considerations about floodplain development were obstructed by the existing policy framework in which the federal government assumed large parts of the costs of flood protection and damage compensation. Under the protective wings of the central state, people continued developing flood-prone areas, and there was no incentive to take the potential costs of flood damage into account in their building choices (White 1945: 210–211). In his thesis, White (1945: 34) therefore called for a "geographical approach" to flood risk management, in which federal floodplain regulations and price incentives would be used to encourage an optimal use of the country's floodplains.

The ideas of Barrows and White proved influential. They were embraced in a number of significant publications on water management in the 1950s and adopted by key scientists in this field (Macdonald et al. 2012). They were also taken to practice. In the Tennessee River basin, economic profits gained by flood protection often did not justify the investments. The Tennessee Valley Authority (TVA), one of Roosevelt's federal agencies, started to experiment with White's approach, which they coined "floodplain management" (Cigler and Burby 1990). The TVA conducted a large number of studies that greatly improved methods for the collection of local flood hazard data and insights about the use of land-use regulations to guide building activities in floodplains. The TVA distributed over 200 research reports between 1950 and 1970, further advocating the idea of floodplain management (Wright 2000: 20). Together with the active advocacy of Barrows and White during public manifestations and governmental meetings, the idea of floodplain management gradually became part of standard discussions in US flood governance (Hinshaw 2006). At the local level, there was resistance to floodplain management. Local states and communities feared that federally imposed land-use regulations would restrict their autonomy and impair their economies. But such concerns were usually suppressed by the active voices of supporters of the floodplain management approach who entered the public debate every time a flood stroke to discredit the excessive federal bailouts the existing policy approach indulged.

4.3.2 Political Acceptance of Flood Insurance (1960–1970)

Growing urbanization in the 1950s and 1960s accelerated floodplain development. More levees were built to protect these growing settlements (Wright 2000). Despite these investments in flood protection, flood damage rose steadily in this period and the federal government increased its expenditures on disaster relief to help victims recover and rebuild after a flood. In response to these events, calls were made to launch a federal insurance scheme for flood damage as a more structural means of flood damage compensation. After the big Mississippi floods of 1927, private insurers had stopped covering flood damage, which left US citizens dependent upon their federal government for damage compensation. While in 1956 a bill was adopted to establish a federal flood insurance scheme, this scheme was never implemented; it was held back because of the difficulty of calculating accurate premiums because of the low probability and catastrophic nature of flood risks. The emergent policy philosophy on floodplain management in the 1960s initiated a renewed interest in federal flood insurance.

Links between insurance and floodplain management were apparent in the work of two task forces on federal flood control that were set up by president Lyndon B. Johnson in the wake of the destructive floods caused by hurricane Betsy in 1965. The first task force was set up within the Federal Department of Housing and Urban Development (HUD). The report criticized existing flood management strategies. Flood protection and federal disaster aid not only created public expectations for federal assistance after floods, but it also undermined rational floodplain management choices that could help to prevent damage in the event of a flood. According to the HUD task force, a more viable alternative would be a "self-help" insurance program that would address the responsibilities of floodplain occupants in mitigating flood damage (HUD 1966: 12). However, the authors realized that the actuarial rates the private insurance industry would have to charge to underwrite the assumed risk would make flood insurance unaffordable for people living in high-risk areas. Therefore, the authors recommended the setup of an insurance program in which the federal government would ensure premiums remained affordable by acting as a reinsurer (lender of last resort) for catastrophic losses and by subsidizing premiums of people living in high-risk areas.

White was contracted to chair the second task force. He brought in James Goddard, a central spokesman of TVA, and both men handpicked the other members of the task force. In their report titled "A Unified National Program for Managing Flood Losses," the floodplain management vision was clearly present. The authors claimed that while the "[u]se of flood plains involving periodic damage from floods is not, in itself, a sign of unwarranted or inefficient development [...] [p]rinciples of national economic efficiency require, however, that the benefits of flood plain occupancy exceed all associated costs" (Task Force on Federal Flood Control Policy 1966: 13–14). Existing policies were criticized for relieving "individual flood plain occupants of responsibility, in a fiscal sense, for the consequences of their actions" while the "general public, by bearing all or a major part of the cost of flood protection works and lessening the individuals' damage costs, further subsidizes their use of the flood plain" (idem.: 15). According to the authors, a more effective approach would be to alter the price signals received by potential floodplain developers, for example in the form of a risk-related "occupancy charge" (idem.: 16). In addition, land-use requirements could be used to prohibit developments in the most hazardous areas. Offering subsidized premiums in a federally supported insurance program as the HUD task force proposed could be counterproductive, White's task force argued, as this would encourage uneconomical developments in the nation's most hazardous floodplains.

When the reports of both task forces were discussed in Congress, the limits of engineered flood protection were emphasized as well. Presented with an alternative to engineering solutions in flood governance, a member of Congress stated that '[e]ngineers admit, as competent as they may be, that they cannot provide complete protection for every flood' (90 Cong. Rec. H 1096 1967). Federal disaster relief programs were criticized. This form of aid was not only given on a "piecemeal disaster-by-disaster basis," but it also provided floodplain occupants with a guarantee for damage compensation without appropriate responsibilities attached to that (90 Cong. Rec. 30791 1967). Insurance was generally accepted as a viable alternative. As one representative put it, insurance is, "in the finest American tradition, helping the citizen to help himself in anticipation of a potential disaster" (90 Cong. Rec. 30807 1967). Besides, that insurance would discourage unwise developments in floodplains and help to reduce flood losses was underlined as an important benefit as well (Subcommittee on Housing 1967: 4–5).

However, worries arose about the affordability of flood insurance. As the HUD report already indicated, insurance premiums were deemed unaffordable for people living in high-risk areas. While White's task force cautioned against the use of subsidized rates for undermining damage mitigation incentives, Congress shared the concerns about affordability with the HUD task force (90 Cong. Rec. H17279 1967). When the National Flood Insurance Program (NFIP) was enacted in 1968, Congress decided to subsidize the insurance premiums of properties located in high-risk areas to ensure their owners had access to reasonably priced damage compensation. Full-risk rates would be charged to new developments or substantially damaged structures in these areas in order to discourage the further development of floodplains. In addition, it was decided to only offer flood insurance in communities that had enacted certain minimum land-use regulations that prohibited building activities in their floodplains. The need for subsidization was thought to be greatest in the first operating years of the program, when a level ranging from 66% to 75% was believed necessary (Subcommittee on Securities 1967a: 81). However, a HUD spokesmen assured Congress that "[a]s the program develops and homeowners in less hazardous zones begin buying insurance, or as new or substantially improved properties come in, the Government's share of the risk will decline" (Subcommittee on Housing 1967: 15), an argument that was also made by the private insurance industry (Subcommittee on Securities 1967a: 81).

The NFIP was set up as a privately run but federally backed insurance scheme. Its implementation was left to private insurers who had organized themselves into a National Flood Insurance Association (NFIA). Premium revenues were collected in a common National Flood Insurance Fund, which was backed by federal resources; in case of catastrophic losses, this fund could borrow from the national treasury to cover claims. Oversight and management responsibilities for the NFIP were placed under the auspices of the HUD department.

With the National Flood Insurance Program, Congress embraced the principles of rational floodplain management. However, it traded off its call for risk-based pricing as an effective flood damage mitigation strategy with concerns about costs this would infer on local-level actors living in high-risk areas. This trade-off resulted in the choice to federally subsidize insurance premiums in these areas.

4.4 The Institutionalization of the US Spatial Approach to Floods

4.4.1 Growing Federal Involvement Under Environmentalism (1970–1980)

In the first operating year of the NFIP, only four communities enrolled and only 20 insurance policies were sold (Platt 1999). In the years thereafter, participation also fell short of expectations. Mapping delays formed an important part of the problem. When the program started out in 1968, HUD was given the gigantic task of mapping all flood hazard areas in the nation. However, data on local flood risks was not readily available and HUD had to collect this while the program was already in operation. Using the standard of a "1:100 flood" already applied by the TVA—which denoted a flood event that had a statistical chance of recurring once every hundred years—HUD used historical flood data to identify flood hazard areas in each community, which it correlated to insurance premiums in Flood Insurance Rate Maps (FIRMs). As communities could only enter the program once HUD had published their FIRM, mapping delays slowed down community participation.

In 1969, an Emergency Phase was introduced that allowed communities to enter the program based on Flood Hazard Boundary Maps, which were less detailed than FIRMs as they only showed floodway boundaries and provided no information about flood depths (Riebau 2000). Because these boundary maps provided little information to support actuarial (risk-based) rate calculations, insurance was offered against subsidized rates and little to no land-use regulations were required from local communities' planning departments. While this Emergency Phase boosted participation in the program, the practice was also criticized, for it not only increased federal subsidization in the program but also undermined the program's original damage mitigation goals.

Despite these struggles, political support for the program was upheld in the 1970s. This was mainly because of the strategic interlinkages between environmental organizations and supporters of the floodplain management approach. The environmentalists' goal of protecting natural areas against economic claims on land coincided well with the purpose of the insurance program to prevent developments in natural floodplains. Influential environmental NGOs such as the Wetlands Organization started to support the NFIP as a tool for environmental planning. In 1975, a new task force on floodplain management was set up at the geography department of Georgia State University. This task force, later renamed the Water Resources Council, included representatives of several federal departments and agencies, including the 1970-established Environmental Protection Agency, HUD, and TVA, whose work knitted together flood safety and environmental goals through land-use management. It was an active group. Members met every six weeks, special working groups more often (Wright 2000). The Water Resources Council produced several reports with guidelines to help communities and individuals meet the NFIP's building and zoning requirements for damage mitigation (US Water Resources Council 1969, 1971).

In policy discussions on the NFIP, the new focus on land-use planning was reflected as well. This was instigated by HUD administrators, who on several occasions proposed to add new rules to the NFIP's regulatory structure to strengthen its land-use planning component. In 1973, HUD for example proposed to require of communities that received disaster assistance after a flood that they would strengthen their building and zoning regulations in affected areas (Subcommittee on Housing and Urban Affairs 1973: 16). HUD also proposed to make flood insurance a mandatory precondition for receiving federal loans and grants for constructions in high-risk areas. These proposals elicited concerns from local communities. As the National Association of Counties pointed out during a hearing in 1973, "the whole economy is built upon retirement, land development, and building, all taking place within the so-called flood plain"; if insurance and land-use planning was required in these areas, they would become "unusable because of the forced zoning elevations", bringing "economic disaster" to those areas (idem.: 85). Congress members also started raising questions about the costs of insurance and rebuilding to NFIP standards at the local level (idem.: 47–49, 69–70). However, HUD continued to stress the importance of land-use planning in the operation of the program. In addition, HUD representatives pointed out that existing structures in high-risk areas would continue to be subsidized. As the administrator of the program at HUD explained: "I don't envision the program becoming self-supporting for many, many years. We are talking about a very, very long-range effect […] [T]his bill was not structured to make the program self-supporting in the near future. It was structured to get construction on the right road" (idem.: 47).

As a result of these discussions, several measures were introduced to strengthen the program's building and zoning requirements (Arnell 1984). A 1973 reform package placed a mandatory purchase requirement for flood insurance on all federally-backed properties in high-risk areas in participating communities. These reforms also required of these communities an assurance that 1:100 floods could be accommodated in their floodplains without water levels rising more than one foot at any point (ASFPM 2004). To help individuals and communities meet NFIP building and zoning standards, the Water Resources Council continuously revised its guidelines for flood damage mitigation, which were adopted as executive orders and brought under the statutory framework of the NFIP (US Water Resources Council 1971, 1979). In this period, the federal government also started to financially assist individuals and communities through loss reduction and damage mitigation grants. To balance these stricter land-use requirements imposed by the NFIP and further increase participation, the Emergency Phase was extended and subsidized rates were lowered several times between 1972 and 1974.

As a result of these measures, community participation increased from 158 in 1971 to 15,000 in 1977 (Platt 1986: 56). However, many insurance policies were sold against subsidized rates. This stirred up a conflict between HUD and the NFIA, the collective of private insurers involved in the program. The NFIA refused to provide HUD access to historical claim data, because of which it remained unclear what percentage of the rate charged by private insurers was actually subsidized by the federal government (NFIA 1977). This led to a breakdown of trust and evenutally a government takeover of the program in 1977 (Subcommittee on Housing 1977). While private insurers continued to be involved in the program as sell and service agents, this decision in effect transformed the NFIP into a fully federal enterprise supervised by HUD. As a consequence of this government takeover, premiums were no longer federally "subsidized", but simply included as "discounted" rates in the general program budget, internally compensated by the revenue generated by received premiums.

As a result of the collaborative efforts of social geographers and environmentalists in the 1970s, land-use planning was highlighted as a policy goal for the NFIP. In policy discussions of the NFIP, this goal was evaluated against the costs involved with stricter land-use requirements in high-risk communities, where large parts of the nation's economic activity clustered. The political trade-off expanded the federal government's role in the program. The NFIP was reformed to impose land-use regulations on local communities, but at the same time, the federal government assumed a greater part of the risk by allowing communities to enter the program against subsidized, and later discounted, rates.

4.4.2 A Liberal Turn (1980–1990)

In the 1980s, under the Reagan Administration (1981–1989), market regulation became the cornerstone of public policymaking. This political change affected the flood insurance program as well. In 1979, management responsibilities for the NFIP were transferred from HUD to the Federal Emergency Management Agency (FEMA), a regulatory agency created by Reagan to improve the coordination and effectiveness of US disaster management. In 1981, Reagan stopped financing the

Water Resources Council, thereby effectively ending an important communication channel between federal government and social geographers that had always been closely involved in the development of the NFIP. While NFIP participation steadily grew, the balance of the National Flood Insurance Fund ranged between negative $5.4 million and negative $323 million between 1972 and 1980, and HUD regularly had to borrow from the federal treasury to pay out damage claims (Pasterick 2000: 191). Resistance against this state-led and state-subsidized program increased.

When the operation of the NFIP was discussed by Congress in the 1980s, it was seen as problematic. Contrary to its intentions, the program had failed to prevent development along the coast and reduce federal disaster assistance. As a senator stressed, "[i]t was through the insurance premium that the program would transfer a substantial portion of flood-related costs borne by taxpayers to those who lived in the flood-prone areas. Through this mechanism, it was envisioned that the program would eventually become self-sustaining. But today, that is not the case. The facts show that the subsidies have been increasing" (Committee on Appropriations 1981: 4). To aid this situation, the Reagan Administration called on FEMA to cut back subsidies to pre-1972 levels (idem.: 2).

In 1981, FEMA itself promulgated a new goal for the NFIP, that is, to be self-supportive for the average loss year. This average loss year was calculated as the annual average of losses experienced in the operational history of the program (i.e., since 1968), and self-supportiveness entailed that the total sum of annual premiums equaled the annual average loss. To meet this goal, FEMA increased subsidized premiums ten times between 1983 and 1995 (Pasterick 2000: 189). Politically pressured to end the Emergency Phase but provided with no federal resources to map local flood hazards, FEMA decided to reissue the boundary maps of communities with a low-growth potential as FIRMs in order to transfer these communities from the subsidized to the regular program (Riebau 2000).

In the 1980s, the NFIP's operation was viewed in a different light. Subsidized premiums were no longer seen as a legitimate trade-off, used to spread the costs and benefits of the program over the nation. Rather, with a political regime that sought to limit federal involvement in the public domain, subsidized rates came to be seen as unjustifiable elements in a market-based program. Several measures were taken to reduce federal subsidies in the program. These measures had their effect. In 1986, the program was "self-supportive" for the first time. While subsidized premiums were never eliminated from the program, in that year premium revenues equaled historical annual losses.

4.4.3 Operational Mode (1990–2005)

Climate change entered the international flood governance discourse in the 1990s. Floods were projected to become more structural and more devastating. National and international flood governance approaches responded to these new flood risks by adapting to, rather than building against, floods (Butler and Pidgeon 2011; Hartzell-

Nichols 2014). It could be expected that the NFIP, with its emphasis on flood risk mitigation, gained importance as a response strategy to climate change. But while the insurance scheme drew the attention of international scholars working on climate change (e.g., De Moel et al. 2009; Aerts and Botzen 2011), domestic interest in the NFIP declined toward the turn of the century. Within FEMA, floods now "competed" with other emergencies, such as international security and terrorism. Moreover, the self-supportive status upheld by FEMA in this period generated the expectation that the program worked fine, that it would be able to cover all claims when a flood would occur.

The NFIP's operation between 1990 and 2005 has been characterized as being in a "maintenance mode" (Riebau 2000: 171). Suffering from a persistent lack of funding, flood maps were only updated when they were challenged by property owners or when new information was provided to FEMA by insurance agents or local governments. While amendments to the program aimed to improve incentives for flood damage mitigation—for example, through the 1990 Community Rating System that provided premium discounts in communities that enacted more stringent land-use regulations than minimally required by the NFIP—in practice such provisions were little utilized and difficult to monitor (Wright 2000).

This relatively calm and steady road in the development of the NFIP was quite suddenly interrupted by Hurricane Katrina, which struck the US Gulf Coast in 2005. For the first time in 19 years, the program lost its self-supportive status. After paying out all claims of the 2005 hurricane season, the program was left with a U$19 billion debt to the national treasury that nobody thought could ever be repaid (Michel-Kerjan et al. 2012). The event elicited a long line of policy discussions on the operation of the NFIP and its feasibility as a flood management strategy altogether, which will be described in the next chapter.

4.5 Expert-Influence in the US Turn to Spatial Measures

This chapter traced back the influence of experts on the transition from a safety to a spatial planning approach in US flood governance. The aim was to analyze to what extent and in what way experts, through their interaction with political actors, shaped the common policy discourse on floods and how this impacted on the recognition and handling of distributive aspects in the democratic decision-making process. In this section, conclusions will be drawn with respect to the role of experts; what experts were involved in the transition to spatial measures in US flood governance and how have they influenced the direction of this shift? The next section specifically looks at the implications of expert-influence on distributive decision-making.

The historical reconstruction provided in this chapter first of all demonstrates that expert-groups were extensively involved in all stages of the evolution of US flood governance in the 20th century. However, in contrast to the Netherlands where engineers formed the dominant source of expertise in flood governance throughout the 20th century (see Chap. 2), in the US different expert-groups were involved at different times. While the first efforts to control floods were led by engineers, geographical experts took over this role in the 1930s and administrative experts guided US flood governance since the 1980s.

Partly, this variety of flood management experts over time can be explained from the organizational efforts of the expert-groups themselves. Drawing on their successful intervention in the construction of the Panama Canal, engineers actively advocated their technical-engineering solutions to floods in the early decades of the 20th century. The public scuffles between "Army Corps" and "civil" engineers in the early decades of the 20th century contributed to the recognition of engineering expertise in the policy domain in general. However, in the 1940s to 1960s, a new group of experts organized itself in flood governance. Affiliated with the influential Chicago school of behavioral sciences, social geographers openly challenged the technical solutions forwarded by engineers and instead presented their "floodplain management" approach as a better (more cost-efficient) alternative.

At the same time, this chapter also highlights the importance of political-contextual factors in understanding the constitution of expert-influence in US flood governance. Engineering solutions never really landed in the political landscape of the US. The engineers' perception of floods as a collective problem that required a federal response clashed with dominant American political values, such as local state autonomy and individual responsibility. When Roosevelt started to use flood control projects as part of his job-creation plans, this motived social geographers to develop an alternative approach to flood governance. Their rational floodplain management alternative corresponded much better to American political values. Social geographers framed floods as a problem of individual location choice for which not the central state but local communities and individuals should bear responsibilities and costs. Especially after the Roosevelt regime, close bonds developed between social geographers and policymakers. Gradually, engineers were "organized out" of the policymaking process whereas social geographers were "organized in". Under this merger of knowledge and interests, a strong policy frame was created that generated the necessary support for the adoption of the geographers' rational floodplain management approach. In 1968, this approach was formalized in the NFIP.

In addition, this chapter also demonstrates that changes in the political context over time opened up space for the emergence of new expert-groups in US flood governance. For example, experts connected to the upcoming environmental movement in the 1970s, whose aim to prevent floodplain development overlapped with the aims of floodplain management, upheld general support for the NFIP during its first cumbersome years. To address low participation and mapping delays, the federal government even expanded its role in the insurance program in this period. However, this increased federal interference in the program backlashed against the rise of political neo-liberalism in the 1980s. Its ideology of market regulation conflicted with the idea of a federally supported flood insurance program. Under the Reagan regime, a new type of expert was "organized" in US flood governance: the operational expert. Management responsibilities for the NFIP were transferred from the federal government to the independent regulatory agency FEMA, and the policy goal shifted from land-use planning to fiscal independence in this period.

Based on these insights, this chapter concludes that in addition to the efforts of expert-groups themselves to get their expertise recognized in US flood governance, dominant political values set boundaries around the type of expertise that was considered relevant for policymaking on floods. Expert-influence in US flood governance can thus best be understood as "contextually embedded" in the larger political context of the US, which limited the space of and set ground rules for expert-groups to organize their expertise in the domain of flood governance. From this contextually embedded understanding of expert-influence, it can also be understood why in certain periods certain expert-groups dominated in US flood governance.

4.6 The Effects of Expert-Influence on Distributive Decision-Making

The previous section analyzed how expert-influence was constituted in 20th century US flood governance. This section reflects on the extent to which and the way in which expert-groups influenced the policy discourse on floods, and through this the distributive aspects of the flood insurance policy that constituted the US spatial planning approach to floods.

The previous section described the influence of social geographers, who stood at the basis of the US spatial planning approach to floods, as being embedded in the larger political context. This embedded character of expert-influence meant that social geographers developed close relationships with political actors in the federal government. The policymaking processes discussed in this chapter show that through the interactions between social geographers and political actors, a "strong" policy frame was developed that provided a coherent story about the problem of floods. In this policy frame, the cause was identified (i.e., the human "encroachment" of floodplains) and a solution was outlined (price incentives for rational floodplain management). In this frame, operational and normative arguments closely interlinked.

Social geographers, through their specific framing of the problem, thus greatly influenced the normative aspects of the US spatial planning approach to floods. However, the analysis in this chapter also demonstrates that the distributive impacts of spatial measures were recognized in the democratic decision-making process. This was so in the decision-making process underlying the adoption of the NFIP in the 1950s and 1960s for example. Here, the findings of the Tennessee Valley Authority, an agency that repeatedly tested the price signals and regulatory standards proposed by geographers, provided insight into the costs spatial measures would confer on individual actors. The results of the Tennessee Valley Authority were actively communicated by social geographers in an effort to demonstrate the practical feasibility and good results of floodplain management. When the NFIP was adopted in 1968, US Congress traded off the goal of sending effective price signals for flood damage mitigation against the goal of affordable flood insurance and decided to subsidize insurance premiums in high-risk areas.

Over time, geographical expertise continued to highlight the local-level distributive effects of spatial measures. The Water Resources Council, which seated many social geographers, recurrently published guidelines for damage mitigation to assist individuals and communities in acting on their newly assigned responsibilities in flood governance. By specifying options for local actors, political actors were able to grasp the practical distributive consequences of US spatial measures. In the 1970s, this facilitated a discussion on a possible conflict of interests between requiring stricter building standards in floodplains, and local communities' freedom and aim to facilitate economic development in these areas. A new trade-off was made, in which the federal governance increased its financial support to the program in return for an expansion of the federal government's right to interfere in local spatial planning.

Up to the 1980s, the interaction between social geographers and political actors facilitated the recognition of the distributive aspects of the US spatial planning policy. As was indicated in the interviews, affordability was always a factor in the policy discussions on the NFIP.[1] However, under neo-liberalism in the 1980s, responsibilities for the NFIP shifted from HUD to the independent regulatory agency FEMA, and a new emphasis was placed on the "operational efficiency" of the NFIP, which in practice meant fiscal independence. As a result of this new policy goal, federal subsidies on insurance premiums and federal financial support for individual damage mitigation measures were cut back. With this, costs and responsibilities for dealing with floods were transferred to local-level actors. However, under the administrative expertise of FEMA, these shifts in costs and responsibilities were justified based on their contribution to the program's operational effectiveness. And while on paper, these policy adjustments contributed to the insurance program's operational effectiveness, in practice, climate change, urbanization and reduced investments in structural flood protection had in fact increased the flood risk, amongst other in New Orleans. When hurricane Katrina revealed the (distributive) implications of these policy choices, they came as a shock to policymakers; experts and political actors alike. The next chapter examines how this situation was dealt with.

References

90 Congressional Records 30693–30848 (November 1, 1967)

90 Congressional Records H1096–H1135 (daily edition January 23, 1967)

90 Congressional Records H17277–H17280 (daily edition June 26, 1967)

Aerts JCJH, Botzen WJW (2011) Climate change impacts on pricing long-term flood insurance: a comprehensive study for the Netherlands. Glob Environ Change 21(3):1045–1060

Arnell NW (1984) Flood hazard management in the United States and the National Flood Insurance Program. Geoforum 15(4):525–542

Arnold JL (1988) The evolution of the 1936 flood control act. Office of History of the United States Army Corps of Engineers, Fort Belvoir

[1]Interview US Government Accountability Office, April 25, 2014, Washington, DC; Interview Resources for the Future, April 22, 2014, Washington, DC.

ASFPM (2004) Background reading for the 2004 Assembly of the Gilbert F. White National Flood Policy Forum 'Reducing flood losses: is the 1% chance (100-year) flood standard sufficient?' Available via ASFPM. http://www.asfpmfoundation.org/forum/2004_Forum_BackgroundPapers.pdf. Accessed 1 Nov 2014

Barry J (1997) Rising tide: the great Mississippi flood of 1927 and how it changed America. Simon & Schuster, New York

Butler C, Pidgeon N (2011) From "flood defence" to "flood risk management": exploring governance, responsibility, and blame. Environ Plann C: Gov Policy 29(3):533–547

Cigler BA, Burby RJ (1990) Floods. In: Waugh WL Jr, Hy RJ (eds) Handbook of emergency management: programs and policies dealing with major hazards and disasters. Greenwood Press, Westport, pp 81–105

Committee on Appropriations (1981) Hearings before a subcommittee of the Committee on Appropriations of the United States Senate, Special Hearing, held on May 6 1981, report no. 79-922. US Government Printing Office, Washington, DC

De Moel H, Van Alphen J, Aerts JCJH (2009) Flood maps in Europe—methods, availability and use. Nat Hazards Earth Syst Sci 9:289–301

Hartzell-Nichols L (2014) Adaptation as precaution. Environ Values 23(2):149–164

Hinshaw RE (2006) Living with nature's extremes—the life of Gilbert Fowler White. Johnson Books, Boulder

HUD (1966) Insurance and other programs for financial assistance to flood victims. Department of Housing and Urban Development, US Government Printing Office, Washington, DC

Layton ET Jr (1986) The revolt of the engineers: social responsibility and the American engineering profession. Johns Hopkins Press, Baltimore

Macdonald M, Chester D, Sangster H, Todd B, Hooke J (2012) The significance of Gilbert F. White's 1945 paper "'Human adjustment to floods' in the development of risk and hazard management". Prog Phys Geogr 36(1):125–133

Michel-Kerjan EO, Lemoyne S, De Forges S, Kunreuther H (2012) Policy tenure under the U.S. National Flood Insurance Program (NFIP). Risk Anal 32(4):644–658

NFIA (1977) Chronology of relationship between FIA and NFIA. National Flood Insurance Association, New York City. http://digitalcollections.library.cmu.edu/awweb/awarchive?type=file&item=544591. Accessed 16 Dec 2014

Pasterick ET (2000) The National Flood Insurance Program: a US approach to flood loss reduction. In: Marsalek J, Watt WE, Zeman E, Sieker F (eds) Flood issues in contemporary water management. Kluwer Academic Publishers, Dordrecht, pp 185–196

Platt R (1986) Flood and man: a geographer's agenda. In: Kates R, Burton I (eds) Geography, resources, and environment: themes from the work of Gilbert F. White, vol II. University of Chicago Press, Chicago and London, pp. 28–68

Platt R (1999) Disasters and democracy: the politics of extreme natural events. Island Press, Washington, DC

Reuss M (1992) Coping with uncertainty: social scientists, engineers, and federal water resources planning. J Nat Resour 32(1):101–135

Riebau MA (2000) The importance of maps for floodplain management and flood insurance. In: Marsalek J, Watt WE, Zeman E, Sieker F (eds) Flood issues in contemporary water management. Kluwer Academic Publishers, Dordrecht, pp 167–176

Simon HA (1985) Charles E. Merriam and the "Chicago School" of Political Science. Edmund James Lecture, October 10, 1985. Available via Digital Commons. http://digitalcollections.library.cmu.edu/awweb/awarchive?type=file&item=34043. Accessed 22 Dec 2014

Subcommittee on Housing (1967) Hearings before the subcommittee on housing of the house committee on banking and currency of the US house of representatives held on August 15 and 18, and September 19, 20 and 21, 1967. Report no. 84-340. US Government Printing Office, Washington, DC

Subcommittee on Housing and Community Development (1977) Hearings before the subcommittee on housing and community development of the committee on banking, finance and urban affairs of the house of representatives held on September 7 and 8, 1977. Report no. 95-895. US Government Printing Office, Washington, DC

Subcommittee on Housing and Urban Affairs (1973) Hearings before the sub-committee on housing and urban affairs of the committee on banking, housing and urban affairs of the US Senate held on June 11, 12, 13, 14 and 15, 1973. Report no. 98-860. US Government Printing Office, Washington, DC

Subcommittee on Securities (1967a) Hearings before the subcommittee on securities of the committee on banking and currency of the US Senate held on June 26, 27, and 28, 1967. Report no. 81-080. US Government Printing Office, Washington, DC

Subcommittee on Securities (1967b) Transcript of proceedings of the subcommittee on securities of the committee on banking and currency of the US Senate held on August 1, 1967. US Government Printing Office, Washington, DC

Tarlock AD (2012) United States flood control policy: the incomplete transition from the illusion of total protection to risk management. Duke Environ Law Policy Forum 23:151–183

Task Force on Federal Flood Control Policy (1966) A unified national program for managing flood losses. House Document No. 465. US Government Printing Office, Washington, DC

US Water Resources Council (1969) Procedures for evaluation of water and related land resource projects. Report to the Water Resources Council by the Special Task Force. US Government Printing Office, Washington, DC

US Water Resources Council (1971) Regulation of flood hazard areas to reduce flood losses, vol 1. US Government Printing Office, Washington, DC

US Water Resources Council (1972) Regulation of flood hazard areas to reduce flood losses, vol 2. US Government Printing Office, Washington, DC

US Water Resources Council (1979) A unified national program for flood plain management. US Government Printing Office, Washington, DC

White GF (1945) Human adjustment to floods. Dissertation, The University of Chicago, The United States

Wright JM (2000) The nation's responses to flood disasters: a historical account. The Association of State Floodplain Managers, Madison

Chapter 5
Policy Developments After Hurricane Katrina: A Case of Overcoming Uncertainty and Value Conflict

5.1 Introduction

As described in the previous chapter, in the US, floods are governed through a spatial planning approach that institutionalized in the second half of the 20th century. Practically embodied in a federal insurance scheme for flood damage, the National Flood Insurance Program (NFIP) uses risk-based pricing to discourage spatial developments in flood-prone areas. As the previous chapter also demonstrated, the goal of sending effective price signals for rational floodplain management has always been traded off against the federal government's duty to ensure access to reasonably priced flood insurance and its goal to facilitate economic growth in local communities. From the start, the federal government subsidized the insurance premiums of people living in high-risk areas and financially supported risk reduction and damage mitigation measures in local communities. Costs and responsibilities for flood prevention and damage compensation have thus always been shared between the federal, the regional and the local level in US flood governance.

In response to the tremendous havoc caused by Hurricane Katrina in 2005, the insurance program was reformed in 2012. These reforms resulted in extraordinary premium increases, which caused a lot of unrest among insurance policy holders and in the housing market. In 2014, a new reform package was adopted to stop the rate increases. This pendulum policy shift raises questions about the extent to which the distributive implications of the 2012 NFIP reforms were recognized and discussed in Congress.

This chapter investigates the role of experts in the policymaking process leading up to the NFIP reforms of 2012. Experts often have a natural authority in risk governance (Beck 1992; Rosanvallon 2008). An often voiced concern is that the rational explanations experts offered by experts in risk governance guide the attention away from the political choices underlying risk management strategies (Ewald 1991). While the extent of expert-involvement has been documented in a number of existing studies (see Chap. 1), it has not been analysed how experts, through their

E. Bergsma, *From Flood Safety to Spatial Management*, Water Governance - Concepts, Methods, and Practice, https://doi.org/10.1007/978-3-319-96716-5_5

rational argumentation, influence risk management strategies. This chapter investigates this "how-question" in the case of the NFIP reforms after Hurricane Katrina.

To analyse how experts influenced the policymaking process underlying the 2012 NFIP reforms, the analytical perspective of framing is used. A policy frame is a coherent "storyline" about a public problem, in which causes are identified and linked to solutions (Rein and Schön 1993; Hajer 2005). By highlighting certain causes of the problem and linking these causes to policy solutions, policy frames help policy actors come to terms with new policy situations. However, certain frames can become dominant in the policymaking process, setting limits around how other policy actors can interpret the policy problem. Policy frames can therefore be powerful political instruments, used by actors to influence public policymaking (Gamson and Modigliani 1989; Benford and Snow 2000). Existing studies have demonstrated that expert-knowledge can provide an important source of justification to policy frames and that experts and political actors often closely collaborate in the production and communication of policy frames (Jasanoff 2004; Hajer 2005; Metze 2014). The influence of such actor coalitions is mediated by larger contextual factors (Entman 1993; Steinberg 1998) and by processes of collective sense-making in response to internal and external changes in the policy field (Dewulf et al. 2009; Van Hulst and Yanow 2016).

Using the analytical perspective of framing, this chapter addresses the following questions: What experts were involved in the policymaking process on the 2012 NFIP reforms, how have these experts influenced the general understanding of the policy problem facing the NFIP after Katrina, and what did this imply for the recognition of the distributive consequences of the policy reforms n US Congress? To analyze these questions, the policymaking process underlying the NFIP reforms is reconstructed based on an analysis of parliamentary records and additional policy and scientific literature. In addition, the results have been checked and fine-tuned through interviews with key stakeholders involved in the policy discussions. The chapter is structured as follows. Section 5.2 provides a short recap of the key principles underlying the NFIP. Section 5.3 reconstructs the policymaking process on NFIP reform after Katrina. The findings are discussed in Sect. 5.4.

5.2 A Recap: The National Flood Insurance Program (NFIP)

When the first colonizers arrived on the US's soil, they settled on the country's most strategically located areas along the Mississippi River. However, life and work in the Mississippi delta was continuously threatened by large-scale flooding disasters. From the beginning of the 20th century onward, important parts of the river were engineered and levees were built to control the Mississippi River. These flood control projects emerged as a local state affair. Over time, the national benefits of these local flood control projects were increasingly recognized and the federal government started to co-fund local levee projects (Arnold 1988). Alongside this expanding federal role in flood protection, the federal government also became increasingly involved in flood damage compensation through disaster relief. Under the protective wings of the state, more people were drawn to riverine and coastal areas. The levee systems protecting these areas did not always hold and flood damages were steadily rising.

The NFIP was adopted in 1968 as a means to curtail the growing federal involvement in flood governance. It was presented as a cost-efficient alternative to flood control. Emphasizing that while "[f]loods are the acts of God, flood damages are the results of the acts of men", insurance premiums were used as price incentives for "rational floodplain management" (HUD 1996: 14). These price incentives should encourage floodplain occupants to take the costs of flood damage into account in their location and building choices, with the overall aim to reduce the value exposed to floods rather than to continue to build against them.

In principle, insurance premiums are calculated based on the flood risk of the area and the flood risk of the property itself, which is determined based on building characteristics such as the elevation of the ground-level floor and the type of materials used in its construction. In addition, flood insurance is only available in communities that enforce certain minimum floodplain management ordinances that restrict building activities in their "100-year" risk areas, a regulatory standard denoting areas prone to floods that have a statistical chance of recurring once every hundred years (Arnell 1984). These areas are identified based on historical flood data and correlated to insurance premiums in Flood Insurance Rate Maps (FIRMs). However, when the program was enacted in 1968, the goal to send effective price signals for rational floodplain development in the form of risk-based insurance premiums was traded off against the general duty of the federal government to help US citizens cope with the impacts of floods. Besides, it was deemed unfair to place all the costs of flood protection and damage compensation on the shoulders of citizens living in coastal and riverine areas, because these areas were of vital importance to the national economy. For these reasons, it was decided to subsidize the premiums of existing properties in 100-year risk areas and to "grandfather in" properties remapped into a 100-year risk area because of changing flood risks by allowing them to continue to pay their old rates. Only new developments are required to pay a full-risk rate. The federal government also financially supports local risk mitigation measures (e.g., the building of levees, the use of water-resistant building materials and communities' buy-outs of repetitive loss structures) to help local communities and property owners lower their insurance premiums. Under the so-called "Community Rating System", premium discounts are offered to communities that have enacted stricter land-use restrictions that minimally required in 100-year risk areas. Insurance is mandated in 100-year risk areas for properties financed through federally backed lenders; outside of these areas, flood insurance is voluntary. An exception is made for levee-protected areas; because property owners already invested in damage mitigation through their state taxes, the purchase requirement is waivered.

If the program does not have enough revenues to pay out claims, the program is allowed to borrow from the national treasury - a loan which should be repaid with interest. From the start, the program had to use this option to ensure damage compensation after large-scale flooding events. Since the 1980s, the Federal Emergency Management Agency (FEMA) administers the program. Its self-proclaimed goal is to direct the program towards financial "self-supportiveness", which means that mean annual premium revenues should be able to cover the costs in the "average loss year". This average loss year is calculated as the annual average of all claims filed in the history of the program (i.e., since 1968). FEMA is authorized to independently adjust rates, but in 1968, Congress did set a 10% cap on the annual rate increases FEMA was allowed to charge. FEMA managed to uphold a self-supportive status from 1986 to 2005, until Hurricane Katrina broke through all records of damage losses suffered since the implementation of the NFIP in 1968.

5.3 A Reconstruction of the Policymaking Process on NFIP Reform After Hurricane Katrina

The damage caused by Hurricane Katrina in 2005 was unprecedented. Insurance costs exceeded the losses ever suffered from a single insured catastrophic event (Michel-Kerjan et al. 2012: 645). While NFIP coverage was fairly low in the affected states—in New Orleans, coverage ranged from 7.3% to 57.7% (idem.: 645)—the NFIP still needed to borrow the exceptional amount of $17 billion from the federal treasury to pay out all claims (Michel-Kerjan 2010: 166). On top of that loan, Congress appropriated an extraordinary amount of $88 billion for disaster relief (Michel-Kerjan et al. 2012: 646).

The NFIP, by statute, includes a sunset provision that puts an expiration date on every reauthorization of the program. While the program was just reauthorized in 2004 and the next reauthorization was only due in 2008, Katrina instigated a series of policy discussions on the functioning of the insurance program. This section reconstructs these Congressional policy discussions by distinguishing different stages in this process.

5.3.1 Different Value Orientations

In the beginning, the NFIP's $17 billion debt raised a lot of questions. It were mostly representatives of coastal states who immediately called for cancelling this debt; if FEMA would be required to repay, insurance premiums would rise substantially and become unaffordable. This would threaten the American dream of homeownership, especially for "low-income folks who have managed against the odds to own their own home", as Congressman Green from the Gulf Coast state of Texas claimed (153 Cong. Rec. H4606-H4607, 2007). Speaking to his colleagues in Congress, he argued that "[w]hen we reauthorize the NFIP again in 2008, we will need to address this [affordability] issue, because we do not want the Flood Insurance Reform Act to become the Low-Income Home-owner Eviction Act" (idem.).

However, waving FEMA's debt directly conflicted with the belief system of conservative Republicans in Congress, who argued that debt forgiveness would go against the original intention of the insurance program. According to Republican Senator Coburn from the inland state of Oklahoma, the "whole purpose behind this bill in the first place […] was to lessen the cost of the American taxpayer in terms of disasters" (154 Cong. Rec. S3947, 2008). Republican conservatives emphasized the moral hazard created by federal disaster relief; if people know they are going to be compensated for their losses, all incentives to take damage mitigation measures are lost. As Senator Coburn pointed out, "[t]he one thing we have not done is we have not asked people in this country, who are in flood-prone areas, to actually be responsible" (idem.). Instead, they called for the elimination of subsidized rates from the program to better address the individual responsibility of floodplain occupants.

For Great Lake states representatives, Katrina challenged the whole distributive logic underlying the insurance program. They argued that they were disproportionately charged for Katrina losses because they too have flood-prone areas where people are required to buy flood insurance even though they rarely claim damage because floods are less frequent and less damaging there than in coastal areas. As a Michigan representative posed, "[e]ssentially, Michigan and other States in the Great Lakes Basin are being forced to subsidize those in other States who are prone to severe weather events. If that's what we are going to do, we should just call it what it is and have a national catastrophic fund as opposed to this national flood insurance fund. In other words, let everybody pay" (156 Cong. Rec. H4689, 2010).

As the policy discussions on the functioning of the NFIP after Katrina started out in Congress, they were characterized by different value orientations on the distributive principles underlying the insurance program. Representatives from hurricane-prone states stressed the need for affordable flood insurance, conservative Republicans highlighted the importance of addressing the individual responsibility of floodplain occupants, and Great Lake state representatives challenged the cost-sharing mechanism underlying the insurance scheme.

5.3.2 Technical Explanations of the Problem

Congress organized several committee hearings to come to a better understanding of the problems facing the insurance program after Katrina. In these meetings, different experts were asked to share their views on the nature of the problem and the way in which they thought these problems could best be addressed.

One of the first experts Congress called on to testify was the NFIP's program director at FEMA. The director pointed out that the NFIP was never set up to cover catastrophic events like Katrina. Premiums were based on average historical losses and Katrina surpassed any event ever witnessed in the history of the NFIP. In addition, the program worked with "subsidized" rates that were not backed by any form of federal funding; these were simply subsumed as "discounted" rates in the general program budget (FEMA 2005: 3). In the director's understanding, the idea always was that in the event of a catastrophic flood, "the Federal treasury would be the means by which that difference would be made up" (Subcommittee on Housing and Community Opportunity 2005: 32). However, the program director also repeatedly underscored that the NFIP does more than simply distribute the costs of flood damage. He pointed out that the program provides important damage mitigation incentives that prevent economic losses from floods but that mapping delays within the program obstructed this objective (FEMA 2006: 6).

The Association of State Floodplain Managers (ASFPM), an organization of local floodplain professionals involved in the NFIP's mapping actions, also highlighted the program's damage mitigation potential (Subcommittee on Housing and Community Opportunity 2007: 30–39). Together with FEMA, this organization continuously stressed that storms are getting bigger and more areas will be at risk to flooding. Levees do not provide sufficient protection against these growing risks, as Katrina demonstrated. However, under NFIP standards, levee-protected areas were exempted from a mandatory purchase requirement. FEMA and the ASFPM called on Congress to extend the NFIP's mandatory purchase requirement from 100-year to 500-year and to levee-protected areas and to strengthen the building and zoning requirements in these areas, for, as FEMA stated, "it doesn't make sense to spend tax dollars to rebuild to outdated standards only to face similar damage when the next storm comes" (Subcommittee on Oversight and Investigations 2007: 17).

The financial industry placed the causes of the problem elsewhere. Just after Katrina, many victims dragged their insurance companies to court, involving the industry in complex judicial trials over the question of who should pay for the damage of an event caused by a combination of technical levee failure, a hurricane, and floods (Manard et al. 2006). Members of the financial industry linked this situation to the structural underfunding of the NFIP, whose historically-oriented rate-setting structure and use of discounted rates prevented the program from building up a sufficient source of revenue to cover all flood-related damage claims.[1] Arguing that that insurance "should operate under the assumption that Hurricane Katrina and indeed the entire 2004/2005 hurricane seasons were not aberrations", members of the financial industry recommended turning the program into a market-based scheme with an "actuarial" (i.e., risk-based) pricing structure (Subcommittee on Oversight and Investigations 2007: 111). This implied that subsidized rates would be eliminated and the program would move faster toward risk-based rates by raising the 10% cap set on annual rate increases.

Proposals to extend the NFIP's mandatory purchase requirement and to implement a risk-based pricing structure raised concerns in the real estate sector. Real estate agents warned Congress members that not everyone would be able to pay a higher insurance premium and that relocating to another area to avoid paying a higher insurance premium is not always an option. When the NFIP's purchase requirement would be extended, families living in 100-year or 500-year areas may find it difficult

[1]Interview National Association of Mutual Insurance Companies, May 13, 2014.

to sell their homes because of the higher insurance costs attached to their properties. The director of National Association of Realtors therefore urged Congress to "strike a balance between ensuring the long-term fiscal viability of the NFIP and avoiding changes that may result in market inequities and housing affordability problems" (Subcommittee on Housing and Community Opportunity 2007: 72).

By connecting flaws in the operational structure of the NFIP to the program's extremely high debt after Katrina, the experts (FEMA, the ASFPM, and the financial industry in particular) consulted in the policy discussions presented a logical and coherent story in which undesirable policy outcomes were linked to operational flaws. However, their arguments also reflected their underlying interests. The ASFPM and FEMA, two actors involved in the NFIP's mapping activities, called for mapping revisions to improve the program's mitigation potential, which would strengthen their own role in the program as well. The financial industry recommended the elimination of discounted rates to restore the NFIP's financial balance, which would bring stability to the insurance market and benefit their operating space. The real estate sector drew attention to affordability problems in relation to possible obstructions on the housing market.

The "operational" interpretation of the problem offered by experts, gradually became dominant in Congress as well. Referring to Katrina, one Congressman for example argued that "history has shown in the last year that we do people no favors by not having an effective flood insurance program, by not helping people prepare; indeed, to the contrary. What we are doing is we are encouraging more people to be in harm's way. […] We are spending billions of dollars that could have been avoided if we had been dealing with an effective flood insurance program" (152 Cong. Rec. H4567-H4568, 2006). Policymakers started to emphasize that the program needed a "tough medicine" to make "the flood insurance program sustainable in the long run" (152 Cong. Rec. H4566, 2006).

The NFIP thus provided the structure through which actors interpreted the problems they experienced after Katrina. The next section describes how the "political" value orientations of politicians and these "technical" problem definitions provided by experts interacted in the policymaking process and resulted in the development of a shared policy frame.

5.3.3 Interaction Between Political and Expert-Arguments: The Development of a Joint Policy Frame

The criticisms forwarded by Great Lake states in which they questioned the distributive principles underlying the insurance program did not fit the operational understanding that became dominant in the policy discussions. The concerns of Great Lake states were generally dismissed by pointing to the insurance rationale that underpins the program. As the ranking member of the House Financial Services Committee, for example, responded, "We are running here a national program. And if it becomes

50 separate State programs or a couple of thousand separate county programs, you lose the insurance principle […]. The government is not a supermarket where you come in and pay for only exactly what you buy off the shelf. There is some joint effort" (152 Cong. Rec. 4613, 2006).

The arguments of conservative Republicans and the financial industry fitted better within the operational understanding of the problem. Referring to the NFIP's outdated rating structure, the financial industry stated that the expectation always was that a large part of the subsidized properties would naturally dissolve from the program, as these properties would be destroyed by floods and their owners would relocate to safer zones where insurance was cheaper. Drawing on this insight, conservative Republicans argued that subsidized rates were never meant to last and now "unintentionally" burdened the American taxpayer (153 Cong. Rec. H10962, 2007). Conservative Republicans and the financial industry therefore met in their call for the implementation of a risk-based pricing structure.

Moderate Republicans started to point to the positive effects of risk-based pricing as well. They generally emphasized how it would increase the self-reliance and local autonomy of communities. A good example of this is provided by Illinois's representative Biggert, who described how the NFIP worked in a town in her district, Tinley Park: "Following remapping in the 1990's, 550 homes in Tinley Park were placed in the floodplain and homeowners would have been forced to pay an extra $1000 per year for flood insurance. However, instead of paying higher insurance rates and leaving homes vulnerable to floods, the residents of Tinley Park took action. They worked on a flood mitigation project, received a revised FEMA approved floodplain map in April of this year, and avoided paying higher insurance premiums" (Subcommittee on Housing and Community Opportunity 2007: 3).

Republican calls for risk-based pricing were, however, criticized by coastal state representatives, the real estate industry, and the ASFPM. The ASFPM continued to urge Congress to "keep in mind that the NFIP has multiple goals, and providing flood insurance that is reasonably priced in order to avoid direct government subsidy of flood damage is an important goal. A number of studies have concluded that if premiums rise too steeply or become too costly, many policyholders will find ways to avoid buying flood insurance. The consequence of having fewer people insured against known risks would be greater reliance on tax-payer funded disaster assistance" (ASFPM 2007: 6). The organization argued that "spreading the risk" by extending the NFIP's mandatory purchase requirement was a better alternative to restoring the financial solvency of the NFIP. By increasing participation in the program, risk-spreading would not only contribute to damage mitigation and loss reduction, but it would also increase insurance coverage and provide the program with a more stable source of income.

However, the ASFPM's suggestion to federally impose a new purchase requirement in 500-year and levee-protected areas was widely opposed. Conservative groups negated this proposal for it would entail risk spreading.[2] Acknowledging that "[a] fundamental tenant of insurance is to spread the risk", a Republican California representative for example argued that "we shouldn't be spreading it to people whose homes will likely never be flooded" (Subcommittee on Oversight and Investigations 2007: 4). But the proposal also met with criticism from coastal state representatives and the real estate industry, for it undermined the American dream of homeownership and the development of the housing market. For Texas's congressman Green, requiring people in 500-year areas to buy insurance also took it one step too far; according to him, supporters of this option often "act like it is the victim's fault when their houses flood, but these critics do not realize that many people did not move to the floodplains, the floodplains moved to them" (152 Cong. Rec. H4606, 2006). The real estate industry was opposed to risk-spreading because it feared new purchase requirements would cause economic hardship on low- and middle-income families living in 500-year and protected areas.

As the risk-spreading option never attained sufficient support, risk-based pricing became the "communicative devise" through which actors talked to each other about the problem and based on which they positioned themselves against each other in the policy discussions on NFIP reform. Fiscally conservative groups and the financial industry underscored the good risk-based rates would do for the taxpayer, and moderate Republicans highlighted its positive effects on local communities' self-reliance. In this context, coastal state representatives, the ASFPM, and the real estate sector started to frame their concerns about affordability directly in relation to risk-based pricing and started talking about the option to slowly phase in risk-based rates to temper rate increases. Congressman Green, for example, argued that spreading out premium increases over time would "make the NFIP more affordable for low-income homeowners, increase participation in the program and decrease the likelihood of a taxpayer bailout in the event of a flood" (153 Cong. Rec. H10987, 2007).

Between 2006 and 2008, different reform bills were produced that moved the program closer to risk-based rates. Because this solution continued to raise concerns about affordability—besides the effects of eliminating subsidized rates from the program, concerns also centered on the implications of requiring FEMA to repay its still-lingering Katrina debt on premium increases—all reform bills struck a balance between debt forgiveness and implementing risk-based rates. There was, however, a considerable difference between how Senate and House bills stroke this balance. Defended as a "fair compromise", House bills required FEMA to repay its debt but only implemented risk-based rates for business properties, second homes, and vacation homes (152 Cong. Rec. H4591, 2006). Senate bills, contrarily, cancelled the debt but in exchange included provisions to eliminate subsidies on a wider range of properties (154 Cong. Rec. S4059, 2008).

[2]Interview Resources for the Future, April 22, 2014, Washington, DC; Interview Association of State Floodplain Managers, April 23, 2014, Washington, DC. Interview FEMA, April 25, 2014, Washington, DC.

5.3.4 Frame Evolution in the Context of Uncertainty

It proved difficult to resolve the differences between House and Senate versions of NFIP reform bills because there was a lot of uncertainty about the impacts of the proposed reform measures on rate increases and the affordability of flood insurance.[3] This was because the NFIP had never really worked with a risk-based pricing structure. In high-risk areas, the use of discounted rates had long obviated the need to develop an accurate risk-based pricing structure, and in low- to medium-risk areas the program always worked with flat rates.[4]

In different committee hearings, Congress members tried to gain a better insight into the impacts of risk-based rates on premium increases. However, none of the committee hearings generated much clarity on these questions, as the conversations between Republican Representative Neugebauer and the NFIP's deputy administrator at FEMA, Mr. Connor, and between Congressman Green and Mr. Minkler of the Independent Insurance Agents and Brokers demonstrate.

> Mr. Neugebauer: Just because of the caps that are on the increases, what would you say, if we passed a bill today that said let's make all premiums, vacations homes, primary homes, let's make them actuarially based, what would be the percentage of increase that most people would be experiencing? Mr. Connor: You know, what I'd like to do is to provide that testimony for the record, because I'd like to go back and just do an analysis on that. (Subcommittee on Housing and Community Opportunity 2007: 19)

> Mr. Green: We've talked about having persons actually pay who are in the targeted areas of floodplains, let them pay the costs of the burden of having repairs or replacement, making it actuarially sound. What will that cost a typical person if we do this? Mr. Minkler: Congressman, I don't have the exact figure. A broad statement would be there will be an increased cost for those. Mr. Green: Do you think it would double what persons are paying now? Could it triple what persons are paying now? Mr. Minkler: I'm sorry, I don't have an answer for you. Mr. Green: Does someone else on the panel have some intelligence to share with us on this? Double, triple? [No response] (idem.: 43)

Coastal state representatives called for postponing the implementation of risk-based rates until its impacts on premium increases were better understood (e.g., 152 Cong. Rec. H4604, 2006). Democratic Senator Landrieu and her Republican colleague Vitter from Louisiana repeatedly insisted on a study into the effects of actuarial rates on the affordability of flood insurance (154 Cong. Rec. S3821, 2008). Amidst this uncertainty, no NFIP reform bill was passed. As a result, the program expired in 2008. Between 2008 and 2011, the program survived on 16 short-term extensions but also lapsed four times in between these extensions.

These program lapses meant that insurance contracts could not be sold and house sales could not be closed in high-risk areas where flood insurance was a mandatory condition for obtaining a mortgage. This happened right in the middle of the economic crisis, which already put a lot of pressure on the housing market. In this situation, Congress was pressured to quickly pass a long-term reauthorization of the program to

---

[3]Interview US Government Accountability Office, April 25, 2014, Washington, DC; Interview National Association of Realtors, April 25, 2014, Washington, DC.

[4]Interview Resources for the Future, April 22, 2014, Washington, DC.

not further distress the housing market. However, its options to act were reduced by an important change in the political landscape. After the 2010 elections, the Republicans seized the majority in the House of Representatives, and their conservative faction blocked any bill not marked as "budget neutral" by the Congressional Budget Office.[5] In effect, this meant they gridlocked every bill that included a provision for debt forgiveness or the prolongation of subsidized rates over time.

Against this background, real estate organizations, who up to that moment had always opposed risk-based pricing out of fear for steep rate increases, started to support reform bills that included risk-based rates to put an end to the policy impasse that burdened their members.[6] Likewise, the ASFPM, who from the beginning onward underscored the importance of embedding the implementation of risk-based rates in a larger framework of risk-spreading to ensure affordability and participation, started to advocate solutions for affordability in means that were disconnected from the insurance program itself to make sure the program would not cease to exist,[7] breaking the organization's role in it down along this path. The ASFPM called for the implementation of a separate means-tested voucher program to make sure flood insurance would remain affordable for low-income families (Subcommittee on Insurance, Housing, and Community Opportunity 2011a: 8-16).

With key critical voices on affordability more or less "side-lined" in the policy discussions on NFIP reform, views on the solution of implementing risk-based rates started to change as well. Increasingly, Congress members started to stress the good risk-based rates would do for restoring an effective operation of the program. Risk-based rates would provide better damage mitigation incentives, reduce future flood losses, and keep premiums affordable in the long run (154 Cong. Rec. S3855, 2008). Even coastal state representatives started to emphasize this "operational" function of risk-based rates. Congressman Green, who had always strongly opposed actuarial rates, stated that "[w]e all know that the flood insurance program plays a critical role in lessening the impact of major flooding disasters; but to make the program more effective, we need greater participation from Americans of all incomes" (153 Cong. Rec. H10987, 2007, see also Congressman Green's statement at 156 Cong. Rec. H5634, 2010). Phasing out risk-based rates came to be seen as a temporary solution to soften the short-term impacts of rate increases that would allow the program to restore its effective operation. Supporting a bill that would move the program closer to risk-based rates, a California representative, for example, argued that phasing in risk-based rates over a period of five years "would address the NFIP's serious financial challenges by directing it back towards fiscal health and self-sustainability" and would also "lower the burden of higher insurance rates" on low-income families (156 Cong. Rec. H5625, 2010).

In 2010 and 2011, the discussion on NFIP reform mainly revolved around how fast to phase out discounted rates (by increasing the 10% cap set on annual rate increases for different kinds of properties) and how fast to phase in risk-based rates

[5]Interview Resources for the Future, April 22, 2014, Washington, DC.

[6]Interview National Association of Realtors, April 25, 2014 (skype interview).

[7]Interview Association of State Floodplain Managers, April 23, 2014, Washington, DC.

for remapped properties (by gradually breaking down the grandfathering provision). At this point, however, some experts started to emphasize the consequences of implementing risk-based rates. At a 2010 committee hearing, the GAO—an independent watchdog organization in US politics that closely followed developments around the NFIP after Katrina—stated that charging risk-based rates to provide better mitigation incentives is one thing, but that such a choice should always be accompanied with "a dialogue about the appropriate role of government in paying for losses for natural catastrophes" (Subcommittee on Housing and Community Opportunity 2010: 13).[8] The ASFPM also prevailed upon Congress to substantiate its policy choices: "Should the NFIP accommodate catastrophic floods [...]? If so, are there realistic, affordable program adaptations that can achieve this objective? And if not, would it not be best to clarify that the program is not expected to cover catastrophic losses?" (Subcommittee on Insurance, Housing, and Community Opportunity 2011a: 7). FEMA started to demarcate its responsibility. It argued that moving toward actuarial rates would lead to higher insurance premiums in many cases: "The tendency has been, as constituents have raised the issue of the fairness of that, that there has been a question of how fast we should move. We will move as fast as Congress directs in allowing us to raise those rates" (Subcommittee on Insurance, Housing, and Community Opportunity 2011b: 3).

However, such reflections came at a time when the reforms, for Congress, were already a done business. As the ASFPM explained during the interview, "when it became clear that full risk rates was really their goal, we started saying, that is good, but if you don't address this affordability thing it is going to turn around and bite you. And that is when we tried to get their attention". But because the reform bill had been included in a larger bill on mobility (the Moving Ahead for Progress in the 21st Century Act of 2012), "in the last six months there was really no opportunity to provide amendments. So even the few people in Congress who realized that this was going to be problem, and wanted to make some changes, it was on a track they couldn't stop" (idem.).[9]

5.3.5 *The 2012 Biggert-Waters Act and Its Implications*

In June 2012, Congress passed a long-term extension of the NFIP in the "Biggert-Waters Act" as included in the Moving Ahead with Progress in the 21st Century Act. This reform package encompassed a range of measures that moved the NFIP closer to a risk-based pricing structure. It called on FEMA to calculate premiums based on catastrophic instead of average loss years and required the organization to repay its debt within 10 years. The cap on annual rate increases was increased from 10 to 20–25%, subsidies for grandfathered properties remapped into a 100-year

[8] Interview US Government Accountability Office, April 25, 2014, Washington, DC.

[9] Citations in this paragraph are from the interview with the Association of State Floodplain Managers, April 23, 2014 (skype interview).

area were phased out over a period of four to five years, and the law also arranged that subsidies would be eliminated upon the sale of a property. When the bill was discussed in Congress, Republican Congresswoman Biggert presented the bill as a necessary measure that "improves the NFIP's financial stability; it will reduce the burden on taxpayers [… and …] help bring certainty to the housing market through a 5-year reauthorization" (158 Cong. Rec. H4621, 2012). According to her Democratic colleague Waters, the Biggert-Waters Act "will make flood insurance more affordable […] and strengthen the financial position of the flood insurance program" (158 Cong. Rec. H4623, 2012).

As referred to in the introduction of this chapter, the implementation of the Biggert-Waters Act had a major impact on rate increases. The complete loss of subsidy upon the sale of a home, which was put in at the last moment probably to make sure the bill would be "budget neutral" and accepted by the Republican conservative fraction, put an almost immediate stop to the sale of subsidized properties, which caused outrage among owners (New York Times 2013). In addition, the actuarial rates that property owners were expected to pay in the end were much larger than anticipated. Stories were reported about homeowners who used to pay $300 and after the reforms were required to pay $8,000 or even $24,000 annually (The Times-Picayune 2013). One year after the adoption of the Biggert-Waters Act, Congress was faced with increasing public outcries about the rate increases, picked up by newspapers and other national media.[10]

At a committee hearing on the Biggert-Waters Act in 2013, Louisiana Senator Vitter, who like many coastal state representatives eventually voted in favor of the Biggert-Waters Act, stated: "We all expected some premiums increases. We knew they were necessary to make the system fiscally sound. But quite frankly, what we have been told to expect since then, is a completely different planet in some cases" (Subcommittee on Economic Policy 2013: online broadcast).[11] His Louisiana colleague Landrieu admitted: "We made a mistake. […] It was not well thought-out. It must be fixed" (idem.). Everyone looked at FEMA to do something to stop the rate increases. However, FEMA explained that its hands were tied: "I need help. FEMA does not have the means to address affordability. The bill was set up to create an actuarially sound system, not an affordable one. I have a specified number of years to increase rates, no means to define, let alone address, affordability" (idem.).

In April 2014, Congress adopted the Homeowner Flood Insurance Affordability Act. This law scaled back annual rate increases to a maximum of 18%, restored subsidized rates for remapped properties, and reimbursed homeowners for paid premiums that exceeded their premium as it would be under the new law. To cover some of these rollbacks, a surcharge of $25-250 was placed on all outstanding insurance policies.

---

[10]All interviews.

[11]Online broadcast available at: http://www.banking.senate.gov/public/index.cfm?FuseAction=Hearings.Hearing&Hearing_ID=46b52a52-4d45-4c47-8ddc-de2f32cd348e [January 4, 2015].

5.4 The Role and Effects of Experts in NFIP Reform

The reconstruction of the policy discussions on NFIP reform laid out in this chapter demonstrates that as the policy discussions on the NFIP started out after Katrina, actors forwarded different views on the nature of the problem. Some of these views were "political" in the sense that they directly addressed distributive aspects of the flood insurance program. These views were primarily, though not exclusively (real estate organizations being the exception), expressed by political actors in US Congress. Experts generally forwarded a more "technical" view, linking the situation that resulted from Katrina to flaws in the program's operational structure. However, these technical explanations rested on normative assumptions as well.

The analysis revealed how a technical problem orientation prevailed in the policy discussions on NFIP reform after Katrina, in which the program's regulatory structure was seen to have become outdated in a new context of climate change. This view not only helped actors to grasp the problems experienced after Katrina, but it also presented them with a road map for action. In order to continue to ensure an effective operation of the program, its regulatory structure needed to be "restored" and "modernized". Of the two alternatives outlined by experts in this respect, only one—that of implementing a risk-based pricing structure to make sure effective damage mitigation incentives would be provided—proved politically viable. The other option to extent the mandatory purchase requirement from 100-year areas to 500-year and levee-protected areas to spread the risk (and costs) over a larger group of people, clashed with political values in American politics.

The reconstruction of the policymaking process also indicates, however, that the development of this technical policy frame did not automatically imply that political views were overshadowed. Rather, political views were redirected toward this technical policy frame when political and expert-frames started to interact. Value judgments were expressed in relation to the solution of risk-based pricing; some supported this solution for emphasizing the individual responsibility of floodplain occupants, whereas others criticized this solution for undermining the provision of affordable flood insurance. These value differences also stirred up a conflict over the question of whether FEMA should be required to repay its Katrina debt; those favoring individual responsibility generally argued against the taxpayer buyouts debt forgiveness would induce, whereas for those stressing the importance of affordability, debt forgiveness was an essential step to prevent steep rate increases.

These value differences blocked policy action; different standpoints on debt forgiveness and risk-based pricing led the program to expire in 2008. This expiration put a stop to house sales in high-risk areas where insurance is a mandatory requirement and therefore placed a huge burden on the housing market, which was already in distress because of the economic crisis. In this situation, critical voices on the solution of implementing risk-based rates started to disappear; most of these critical arguments had been voiced by actors who formed part of the institutional arrangement in US flood governance and who as such were dependent upon the continued existence of the insurance program in the future. Putting the program back on track became an important new goal for all policy actors, experts and political actors alike.

From this shared goal, an "operational" understanding of the solution of risk-based rates emerged. The idea was that an effectively operating insurance program would minimize flood losses in the long-term, which would keep premiums affordable and address the responsibility of floodplain occupants. The phasing-in of risk-based insurance premiums over a number of years came to be seen as an adequate means to deal with affordability concerns, although it was unclear what these increases would precisely entail. The fact that different political valued merged in this policy frame (i.e. the frame was "inclusive", see Sect. 1.3.2), created the necessary common ground for a long-term reauthorization of the NFIP in the Biggert-Waters Act.

Interestingly, framing scholars have already drawn attention to the influence of such "operational" frames in public policymaking, especially in the domain of risk governance (Vogel 2008; Dunlop 2010; Rayner 2012). As Rayner (2012: 111–120) and Dunlop (2010) have argued in particular, operational policy arguments can help policymakers deal with the uncertain and conflictive circumstances hamper political action in risk governance because they offer a way to reduce value conflicts and uncertainty to technical certainties about the operation of a risk management policy. The possible drawbacks of such operational frames have been outlined as well. Operational arguments generally offer a simplified and future-orientated outlook on a policy solution that emphasizes how the solution would work under ideal circumstances (Enserink et al. 2013; Van der Steen and Van Twist 2013). In doing so, attention is drawn away from the implications of these policy solutions in practice. Consequently, the distributive impacts of a policy solution may not be recognized and value trade-offs behind policy choices may not be sufficiently legitimized in the policymaking process (Vogel 2008; Bressers et al. 2012; Anderson 2013).

This explanation about the role and impact of operational policy frames on policymaking processes seems to provide a good characterization of the policymaking process on NFIP reform after Katrina. Rather than the sole product of expert-actors involved in policy discussions on the NFIP, the focus on an effectively operating insurance program can be seen as a "collective framing reaction" to a new problem that emerged from 2008 onward in which value conflicts and uncertainty hampered policy action in the context of an economic crisis that demanded an urgent response. The operationalized view on risk-based rates helped policy actors deal with this situation by uniting different value orientations in a future outlook on the positive effects created by having an effectively operating insurance program. However, the unrest instigated by the implementation of risk-based rates in the 2012 reforms indicates that the distributive impacts of this solution were not sufficiently recognized in the policymaking process.

References

152(82) Congressional Records H4502–H4503 (daily edition June 22, 2006)
152(85) Congressional Records H4565–H4618 (daily edition June 27, 2006)
153(145) Congressional Records H10947–H11009 (daily edition June 27, 2007)
154(75) Congressional Records S3844-S3932 (daily edition, May 7, 2008)
154(76) Congressional Records S3937-S3976 (daily edition, May 8, 2008)
156(105) Congressional Records H5621–H5658 (daily edition July 15, 2010)
156(95) Congressional Records H4686–H4732 (daily edition June 23, 2010)
158(100) Congressional Records H4610–H4636 (daily edition June 29, 2012)
Anderson MW (2013) Intergenerational bargains: negotiating our debts to the past and our obligations to the future. Futures 54:43–52
Arnell NW (1984) Flood hazard management in the United States and the national flood insurance program. Geoforum 15(4):525–542
Arnold JL (1988) The Evolution of the 1936 Flood Control Act. Office of History of the United States Army Corps of Engineers, Fort Belvoir
ASFPM (2007) Testimony prepared for the hearing of the U.S. Senate Committee on Banking, Housing, and Urban Affairs, 2 Oct 2007. Available via the US Senate. http://www.banking.senate.gov/public/index.cfm?FuseAction=Hearings.Hearing&Hearing_ID=c82cb53d-35a6-42c7-ac25-dc3d7e84db3b. Accessed 4 Jan 2015
Beck U (1992) Risk society—towards a new modernity. SAGE publications, London
Benford RD, Snow DA (2000) Framing processes and social movements: an overview and assessment. Ann Rev Sociol 26:611–639
Bressers N, Van Twist M, Ten Heuvelhof E (2012) Exploring the temporal dimension in policy evaluation studies. Policy Sci 46(1):23–37
Dewulf A, Gray B, Putnam L, Lewicki R, Aarts MNC, Bouwen R, van Woerkum CMJ (2009) Disentangling approaches to framing in conflict and negotiation research: a meta-paradigmatic perspective. Human Relat 62(2):155–193
Dunlop CA (2010) The temporal dimension of knowledge and the limits of policy appraisal: biofuels policy in the UK. Policy Sci 43(4):343–363
Enserink B, Kwakkel JH, Veenman S (2013) Coping with uncertainty in climate policy making: (Mis)understanding scenario studies. Futures 53:1–12
Entman RM (1993) Framing: towards clarification of a fractured paradigm. J Commun 43(4):51–58
Ewald F (1991) Insurance and risk. In: Burchell G, Gordon C, Miller P (eds) The Foucault effect: studies in governmentality. Harvester Wheatsheaf, London, pp 197–210
FEMA (2006) Testimony prepared for the hearing of the U.S. Senate Committee on Banking, Housing, and Urban Affairs, 25 Jan 2006. Available via the US Senate. http://www.banking.senate.gov/public/index.cfm?FuseAction=Hearings.Hearing&Hearing_ID=60b45330-2cdb-4047-8628-b9ffaa17f5a7. Accessed 4 Jan 2015
Gamson WA, Modigliani A (1989) Media discourse and public opinion on nuclear power: a constructionist approach. Am J Sociol 95(1):1–37
Hajer M (2005) Coalitions, practices, and meaning in environmental politics: from acid rain to BSE. In: Howarth D, Torfing J (eds) Discourse theory in European politics. Palgrave Macmillan, Basingstoke, pp 297–315
HUD (1966) Insurance and other programs for financial assistance to flood victims. Department of housing and urban development. US government printing office: Washington, DC
Jasanoff S (ed) (2004) States of Knowledge: the co-production of science and social order. Routledge, London
Manard JP Jr, O'Hara P, Blackwood KR (2006) Katrina's Tort Litigation: an imperfect storm. Nat Resour Environ 20(4):31–77
Metze T (2014) Fracking the debate: frame shifts and boundary work in Dutch Decision making on shale gas. J Environ Policy & Plann 19(1):35–52

Michel-Kerjan EO (2010) Catastrophe economics: the national flood insurance program. J Econ Perspect 24(4):165–186

Michel-Kerjan EO, Lemoyne S, De Forges S, Kunreuther H (2012) Policy tenure under the U.S. National flood insurance program (NFIP). Risk Anal 32(4):644–658

New York Times (2013) Cost of flood insurance rises, along with worries. New York Times, online article, 12 Oct 2013. Available via New York Times. http://www.nytimes.com/2013/10/13/us/cost-of-flood-insurance-rises-along-with-worries.html?_r=0 Accessed 11 Feb 2016

Rayner S (2012) Uncomfortable knowledge: the social construction of ignorance in science and environmental policy discourses. Econ Soc 41(1):107–125

Rein M, Schön D (1993) Reframing policy discourse. In: Fischer F, Forester J (eds) The argumentative turn in policy analysis and planning. Duke University Press, Durham, pp 145–166

Rosanvallon P (2008, trans: Goldhammer A) Counter-democracy—politics in an age of distrust. Cambridge University Press, Cambridge

Steinberg MW (1998) Tilting the frame: considerations on collective action framing from a discursive turn. Theory and Soc 27(6):845–872

Subcommittee on Economic Policy (2013) Hearing before the subcommittee on economic policy of the committee on banking, housing and urban affairs of the U.S. Senate held on 18 Sept 2013. Available via the Senate Banking Committee. http://www.banking.senate.gov/public/index.cfm?FuseAction=Hearings.Hearing&Hearing_ID=46b52a52-4d45-4c47-8ddc-de2f32cd348e. Accessed 4 Jan 2015

Subcommittee on Housing and Community Opportunity (2005) Hearing before the subcommittee on housing and community opportunity of the committee on financial services of the U.S. House of Representatives held on 20 Oct 2005. Report no. 109-60. U.S. Government Printing Office, Washington, DC

Subcommittee on Housing and Community Opportunity (2007) Hearing before the subcommittee on housing and community opportunity of the committee on financial services of the U.S. House of Representatives, 12 June 2007. Report no. 110-38. U.S. Government Printing Office, Washington, DC

Subcommittee On Housing and Community Opportunity (2010) Hearing before the subcommittee on housing and community opportunity of the committee on financial services of the U.S. House of Representatives held on 21 April 2010. Report no. 111-126. U.S. Government Printing Office, Washington, DC

Subcommittee on Insurance, Housing and Community Opportunity (2011a) Hearing before the subcommittee on insurance, housing and community opportunity of the committee on financial services of the U.S. House of Representatives held on 11 March 2011. Report no. 112-16. U.S. Government Printing Office, Washington, DC

Subcommittee on Insurance, Housing and Community Opportunity (2011b) Hearing before the subcommittee on insurance, housing and community opportunity of the committee on financial services of the U.S. House of Representatives held on 11 Mar 2011. Report no. 112-23. U.S. Government Printing Office, Washington, DC

Subcommittee on Oversight and Investigations (2007) Hearing before the subcommittee on oversight and investigations of the committee on financial services of the U.S. House of Representatives held on 28 Feb 2007. Report no. 110-7. U.S. Government Printing Office, Washington, DC

Taxpayers for Commonsense (2013) Testimony prepared for the hearing of the subcommittee on economic policy of the committee on banking, housing and urban affairs of the U.S. Senate held on 18 Sept 2013. Available online via the Senate Banking Committee. http://www.banking.senate.gov/public/_cache/files/755ecc64-ff84-406e-89ae-c8dccfa05726/23C6AE00CC53D934925l1CC744028B5E.ellistestimony91813ep.pdf. Accessed 4 Jan 2015

The Times-Picayune (2013) Flood insurance: whose premiums might go up, and when? Nola.com/The Times-Picayune, online article 13 Aug 2013. Available via NOLA. http://www.nola.com/politics/index.ssf/2013/10/flood_insurance_whose_premiums.html. Accessed 11 Feb 2016

Van der Steen MA, Van Twist MJM (2013) Foresight and long-term policy-making: an analysis of anticipatory boundary work in policy organizations in The Netherlands. Futures 54:33–42

Van Hulst M, Yanow D (2016) From policy "Frames" to "Framing": theorizing a more dynamic, political approach. Am Rev Public Adm 46(1):92–112

Vogel KM (2008) 'Iraqi Winnebagos™ of death': imagined and realized futures of US bioweapons threat assessments. Sci Public Policy 35(8):561–573

White GF (1945) Human adjustment to floods. Dissertation, The University of Chicago, The United States

Chapter 6
A Comparative Analysis of Expert-Influence in Dutch and US Flood Governance

6.1 Expert-Influence in Flood Governance: Blessing or Curse?

Floods are one of the most serious and deadliest risks posed to society. Of all natural catastrophes, floods already cause most insured damages worldwide. Due to climate change and the urbanization of deltas, more losses are expected in the years to come (Swiss Re 2012: 6; Bouwer 2010: 105). At the same time, uncertainties about the impacts of climate change, population variations in delta regions, and the effects of adaptation measures make it difficult to capture future flood risks. This book has dealt with the difficulty of finding effective and legitimate management strategies to control mounting flood risks in a context characterized by uncertainty while decisions can have huge implications for how costs and responsibilities for flood protection are divided.

In the governance of risks such as floods, experts often play a large role. They analyze the causal relationships underlying risk events and calculate risk probabilities, providing a certain amount of predictability to flood risks and their causes. In doing so, experts provide policymakers with tools to devise effective management strategies in flood governance. Like all public policy solutions, these expert-based solutions for flood risks are grounded in deeper layers of understanding, where presumptions on how the world operates and basic norms and values shape the understanding of flood risks and their solutions (Fischer 1995). From a democratic perspective, it is crucial that these normative assumptions are highlighted, discussed and evaluated in an open dialogue (Habermas 1996). This not only enhances the legitimacy of policy choices but also contributes to their implementation because the roles and responsibilities of different actors have been clarified in the policymaking process. Thus, an important challenge in flood governance lies in finding a good balance between expert and democratic forms of decision-making.

The often rational character of expert-explanations and the high-reliance on expert-knowledge in risk governance has given rise to growing concerns about the

E. Bergsma, *From Flood Safety to Spatial Management*, Water Governance - Concepts, Methods, and Practice, https://doi.org/10.1007/978-3-319-96716-5_6

extent to which standard democratic procedures are able to filter out the normative assumptions and implications of expert-solutions in decision-making on risks. However, while many studies have demonstrated *that* experts influence public policy discussions in risk governance, the actual consequences of expert-influence on the quality of democratic deliberations have been analyzed much less (Boswell 2009; Beveridge 2012; Lundin and Öberg 2014; Spruijt et al. 2014). This book, therefore, set out to empirically analyze the effects of expert-influence on distributive decision-making in flood governance.

Empirically, the focus lay on the shift from traditional safety measures that aim to prevent floods (e.g., levees, floodwalls and other structural defense works) to a new category of spatial measures where the aim is to minimize the impacts of a flood event by reducing exposure to flood risks (Botzen and Van den Bergh 2008; Neuvel and Van der Knaap 2010; Fournier et al. 2016; Hegger et al. 2016). In a safety approach, responsibilities for flood protection, emergency management, and damage compensation predominantly lie at the (central) governmental level (Pahl-Wostl 2007). For spatial measures, costs and responsibilities tend to be decentralized to the regional and local level to incentivize risk reduction and damage mitigation measures at these levels (Meijerink and Dicke 2008; Butler and Pidgeon 2011; Bergsma et al. 2012; Penning-Rowsell and Pardoe 2012; Paudel et al. 2015). Many flood-prone countries are currently implementing spatial measures as a response strategy to climate change. This book has analyzed the implementation of spatial measures in two "extreme" cases: The Netherlands, where the safety paradigm has strong historical roots, and the US, where a spatial approach to floods was already implemented in the mid-20th century. The research question guiding the analysis was: **What experts were involved in the shift from safety to spatial measures in Dutch and US flood governance and how have these experts influenced the distributive decision-making process underlying this shift**?

The analysis was structured by a conceptual framework (see Chap. 1) that is grounded in theories of institutional change and gradual institutional adaptation (Hall 1993; Hall and Taylor 1996; Mahoney and Thelen 2010; Gray et al. 2015). Theories of gradual institutional adaptation hold that policy institutions—which refer to the "taken for granted or legitimate models" that structure how a policy problem is perceived and acted upon (Clemens and Cook 1999: 444; Hajer 2005)—do not suddenly change as a result of critical shifts in agency and context, but rather evolve gradually in response to the political agency that is constantly asserted in public policymaking and incremental changes that evolve in the context of the policy field. In this view, flood governance institutions, including their distributive underpinnings, do not suddenly "turn" spatial but move gradually in that direction in response to agency and contextual developments in daily public policymaking.

As a further conceptualization of gradual institutional change, the Policy Arrangements Framework (PAF) was used (Van Tatenhove et al. 2000; Arts and Van Tatenhove 2004; Arts et al. 2006). The PAF specifies four dimensions of policy institutions. First, the content of policy institutions is shaped by a discourse dimension that encapsulates the dominant way of thinking about a policy problem. Then, there are three organizational dimensions that structure how the problem is dealt with.

There is an actor dimension that encompasses the composition of actors standardly involved in public policymaking, a resources dimension that covers the allocation of power and resources to different actors in the policy field, and a rules dimension that specifies the rules and procedures under which policy actors interact. The distribution of costs and responsibilities for flood governance is shaped by the institutional structure of the policy field. Basic allocation principles are formed in the policy discourse, which are further translated into the actor (who), resources (how much) and rules dimensions (by what means). Policy arrangements provide stability; however, the PAF explicitly notes that this structure is only temporal; it is under constant pressure by changing contextual and agency factors in public policymaking, resulting in gradual institutional adaptations at each dimension and within the framework as a whole.

The aim of this book was to analyze how experts influenced the distributive decision-making process underlying the institutional shift towards spatial measures in Dutch and US flood governance. Framing theory was used to better understand agency in these processes. Framing theory poses that actors use storylines (called "policy frames") in which certain causes underlying a policy problem are highlighted and linked to policy solutions, simplifying a complex situation to a structured policy problem that represents their views and interests (Rein and Schön 1993; Hajer 2005). Recent framing accounts highlight the close relationship between political agency, context and collective meaning-giving in the framing process, leading to the production of shared and contextually embedded policy frames (Entman 1993; Steinberg 1998; Dewulf et al. 2004, 2007, 2009; Dewulf and Bouwen 2012: 169; Van Hulst and Yanow 2016).

Taken together, this conceptual framework implies that experts can—deliberately or unconsciously—influence institutional adaptation in flood governance. They do so by offering explanations on (changing) flood risks in the policymaking process, influencing the policy discourse on floods and through this, the organization of flood governance as well—the distribution of costs and benefits being part of that. However, this influence should always be understood against the background of the specific (historical) institutional context in which expert influence is constituted, developments external to the policy field such as climate change, the political agency asserted by actor-groups of which experts may be part, and the collective meaning-giving of policy actors who try to come to grips with a complex policy situation. Figure 6.1 depicts the relationship between the PAF and framing.

To analyze the implications of expert-influence on the extent to which and ways in which the distributive implications of spatial measures were recognized and discussed in the decision-making process underlying their implementation, the research question was split into the following sub-questions:

1. How has the shift from safety to spatial measures in flood governance been framed in the Netherlands and the US?
2. To what extent and how has the policy arrangement underlying Dutch and US flood governance, including its distributive aspects, changed as a result of these framings?

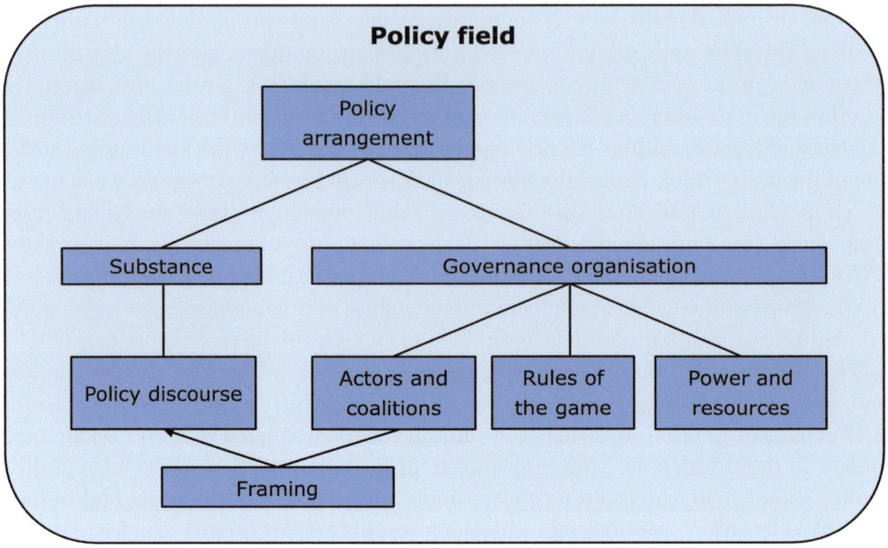

Fig. 6.1 Framing and the policy arrangements framework (adapted from Arts et al. 2006: 99)

3. Which experts were involved in the policymaking process on spatial measures in Dutch and US flood governance and how have they influenced the framing and institutional adaptation processes in both countries?
4. What did this influence of experts imply for the recognition of the distributive implications of spatial measures in the democratic decision-making process?

To fully comprehend the changes underlying the shift from safety to spatial measures in Dutch and US flood governance, the national cases covered over a century of policymaking on floods, from about 1900 up to 2015. This way, the role of experts in (framing) traditional safety approaches to floods could be grasped, based on which changes in the constitution and impacts of expert-influence in Dutch and US flood governance during the shift to spatial measures could be investigated. In this concluding chapter, the findings of the national cases are comparatively analyzed.

6.2 The Shift to Spatial Measures in Dutch and US Flood Governance

This section focuses on the institutional adaptations that resulted from the shift from safety to spatial measures in Dutch and US flood governance. These institutional adaptations will be described in terms of the conceptual framework; this section reflects on the framing that underlay the implementation of spatial measures in Dutch and US flood governance (sub-question 1) and on the way in which these framings altered the policy arrangements in both countries (sub-question 2).

The cases presented in this book show that different institutional paths were followed in the shift to spatial measures in Dutch and US flood governance. They also reveal that a cause of this difference lies in how the shift to spatial planning was framed.

In both cases, a safety approach to floods emerged at the turn of the 19th into the 20th century (see Chaps. 2 and 4). This approach relied on a policy discourse in which floods were seen as an external risk that posed a threat to society as a whole, which was to be dealt with through collective state responses. Policymaking on floods under this safety discourse relied on a combination of engineering and economic expertise, which was used to calculate flood chances and impacts, and to design "optimal" flood defenses for which the costs of investments were traded off against the protection provided by these defenses.

This discourse was reflected in the organizational dimensions of the policy arrangement in Dutch and US flood governance. At the actor dimension, engineers and economist played an important role in public policymaking on floods. This, for example, became apparent in the large involvement of the Zuiderzee Society in the policymaking process on the Zuiderzee Works, a pioneering plan developed by engineers, national economists, business leaders and members of the Dutch landed gentry—organized into the Zuiderzee Society—to close of the (now former) inlet of the at times devastating Zuider Sea with a 32-km long dam (the "Afsluitdijk") to turn this water body into a large freshwater lake with rich plots of newly reclaimed agricultural land. In the US, it was the Army Corps of Engineers that was actively involved in the planning and implementation of river streamlining and flood-control projects and along the Mississippi river in the first decades of the 20th century. In both countries, engineers gained access to important positions in public policy- and decision-making on floods, endowing them with resources to further influence policymaking in this domain. Under this safety approach, the costs of flood control, emergency management and damage compensation were largely borne by the national government, although developments in the US were characterized by a continuous struggle over the level of federal funding vis-à-vis that of "local contributions" of states, and rising federal payments on disaster relief and damage compensation met with increased opposition as well.

In the Netherlands, the safety approach grew stronger over the course of the 20th century. Under this approach, the "Delta Works"—another icon of Dutch flood protection—were constructed in the 1950s to 1960s to protect the low-lying southwestern delta of the country against flooding. Spatial measures were included in the 1990s as part of the safety approach on floods. They were presented by engineers as a more efficient way to deal with the increasing flood risks that faced the country at the end of the 20th century. They were included in the existing policy discourse as an alternative means of providing flood protection against lower costs. Since the 1990s, the Dutch safety approach has been conceptualized as being multi-layered; in the first "safety" layer, the national government still provides protection against floods through structural defense works like levees and floodwalls, but this layer is followed by a second "spatial planning" layer which aims to reduce the exposure to flood risks and a third "emergency management" layer which focuses on increasing the Dutch disaster management capacity in the event of a large-scale flooding disaster. Gradually, the Dutch policy discourse on floods has incorporated individual location and building choices as one of the causes underlying flood risks.

The organization of the policy domain largely remained the same with the implementation of spatial measures (see Chap. 3). At the actor dimension, engineers, together with national economists, still determined what type and level of flood protection were most efficient at which locations, and costs and responsibilities were still largely borne by the national government. However, the analysis laid out in this book also showed that this allocation of costs and responsibilities is currently shifting. As the implementation of spatial measures in Dutch flood governance started out in the 1990s with projects such as Room for the River, the expropriations needed to create space for water were paid for by the national treasury while the costs of disaster relief and damage compensation were also still shouldered by the national government. However, recent policy developments under the Second Delta Program indicate that the national government is taking a step back in some cases of flood protection. While with this program, the national government is increasing its grip on flood protection in high-risk areas, the program explicitly states that people in low- to medium risk areas are expected to also adapt to flood risks themselves through a "second layer" of spatial solutions to flood protection. While in cases of large-scale flood events the national government still has a legal duty to compensate damage, it is not clear what new emphasis on individual responsibility in low- to medium-risk areas means for the distribution of costs and responsibilities for flood protection and small-scale damage compensation in these areas.

In contrast to the Netherlands, the implementation of spatial measures in US flood governance represented a sharp break with the existing safety approach that institutionalized over the first half of the 20th century. Here, safety measures were part of a new policy discourse on floods, in which the root cause of the problem was no longer placed in external factors but in anthropogenic factors (human location and building choices in floodplains) that increased the vulnerability to flooding (see Chap. 5). This policy discourse was forwarded by social geographers associated with the Chicago School of Behavioral Sciences, who gradually gained access to important positions of policy—and decision-making on floods. Rather than keep investing in levees that only facilitate new developments in flood-prone areas, these geographers favored policy solutions that target the human "encroachment" of floodplains, who, from their behavioral background, preferred the use of price signals to influence human behavior. Practically embodied in a National Flood Insurance Program (NFIP) adopted in 1968, insurance premiums were used to incentivize people to take flood risks into account in their location and building choices.

With this shift to a spatial approach in US flood governance, the organization of this policy field also changed. From the 1950s to the 1970s, social geographers occupied pivotal positions on policy advisory committees. Property insurers were made responsible for the implementation of the insurance program, watched over by the Department of Housing and Urban Development (HUD) of the federal government. In high-risk areas (100-year flood zones), communities could enroll in the insurance program by enacting certain land-use and building codes that prohibited further developments in floodplains, and property- and homeowners in these areas were mandated to take out flood insurance. However, the premiums of existing properties in 100-year areas were subsidized by the federal government, which also set a 10% cap on annual premium increases insurers were allowed to charge. Thus, as a result of the implementation of a spatial policy program in US flood governance, the constellation of actors involved in flood governance, the resources provided to them and the rules under which they operate changed substantially. Policy institutions transformed from a federally-led safety approach into a new, spatial governance scheme in which costs and responsibilities for flood protection, disaster recovery and damage compensation were shared between public (the federal government and local municipalities) and private (insurers and individuals) actors.

At least, that was the plan. In practice, the federal government continued to shoulder large parts of the costs in flood governance, through subsidizing premiums and by providing federal disaster relief after a flood. The idea always was that the price signals included in the insurance program would lead people away from flood-prone areas. However, the continued financial support of the federal government to the program was increasingly seen to undermine the effectiveness of these price signals. In the 1980s, the administration of the insurance program was handed over from HUD to the newly created Federal Emergency Management Agency (FEMA). At FEMA, the policy program was overseen by regulators, who were briefed to restore the financial solvency of the program. They did so by gradually phasing out federal subsidies on insurance premiums up to the level that annual revenues matched annual historical costs. In the 1980s, the policy discourse underlying US flood governance transformed into an administrative one, focused on making the insurance program financially sound. This policy discourse brought changes to the organization of the policy field; ties with social geographers broke down with the transference of management and oversight tasks to FEMA, giving influence to FEMA's policy administrators instead. The phasing out of federal subsidies under FEMA's rule only marginally changed the distribution of costs and responsibilities in US flood governance; they just made sure revenues covered expected costs.

Hurricane Katrina demonstrated that there was a huge difference between the damage expected based on the historical flood damage data of FEMA and the actuality of this event. The description of the policymaking process after Katrina revealed a complex policy situation in which collective sense-making and agency interacted in a context of climate change. It led to the implementation of the Biggert-Waters Act in 2012, which phased out subsidies and implemented risk-based rates based on expected future costs of climate change at such a fast speed that costs became unbearable to individual property owners. In 2014, parts of the Biggert-Waters Act were reversed in the Homeowners Flood Insurance Affordability Act, leaving the institutional structure of the policy field that developed in the 1980s to 2000s mainly intact. This means that the policy discourse is mainly focused on maintaining the financial solvency of the insurance program from the understanding that this will provide the most effective price signals for damage mitigation, a status managed by administrative experts at FEMA. The federal disaster relief will be used to cover emergency costs after large-scale flooding events, costs for flood protection and "normal" damage compensation increasingly shift—though now at a more gradual pace—to individuals.

Comparing institutional adaptations in the Netherlands and the US, what stands out is that in the Netherlands safety measures were incorporated in the institutionalized safety approach to floods whereas, in the US, the implementation of spatial measures comprised an institutional shift from a safety to a spatial approach in flood governance, which only grew stronger over the years. In both countries, though at different times and at a very different speed, the shift to spatial measures has resulted in a gradual shift of responsibilities from the national government to local communities and individuals, particularly in the domains of flood protection and damage compensation.

6.3 The Influence of Experts on the Shift to Spatial Measures

The previous section has shown that the implementation of spatial measures gradually changed the institutional framework in Dutch and US flood governance. This section will pay closer attention to the role of experts in producing and guiding these institutional shifts. It thus focuses on the third sub-question, which looks at which experts were involved in the shift to spatial measures in Dutch and US flood governance and how these experts actually influenced the processes of institutional adaptation in both countries.

As the first centralized flood governance strategies developed in the Netherlands and the US at the end of the 19th century, they were based on the expertise available at that time: engineering expertise. As explained in the previous section, this expertise brought a policy discourse in which floods were seen as an external, collective risk that should be dealt with by central state responses. However, as was also discussed in the previous section, the body of expertise underlying Dutch and US flood governance also changed over time, changing the policy discourse and the organization of flood governance in both countries. What can be said about the influence of experts on these institutional paths?

The case studies confirm that Dutch and US policymakers were indeed very dependent upon experts. Experts handed the storylines that policy actors needed to act. This dependency provided experts with an important source of influence in the policy-making process. At the same time, the case studies also reveal that policymakers set limits around the influence of these experts. Both in Dutch and in US flood governance, experts had to tailor their knowledge to prevalent political goals and values to get their expertise recognized by policymakers. Therefore, the influence of experts on policy developments in Dutch and US flood governance can best be understood as being "contextually embedded".

From this contextually embedded understanding of expert influence, differences in the institutionalization of spatial measures between both countries can also be better understood. While the safety approach developed by engineers fit the political landscape of the Netherlands in the late 1900s where the rise of progressive liberalism pushed for larger state involvement in Dutch flood governance, it never took full shape in the US, where values of individual responsibility and local state autonomy were prioritized over large-scale state involvement. The safety approach institutionalized in Dutch flood governance, but, in the US, policymakers embraced the alternative floodplain management approach developed by social geographers in the 1940s. In this process, new structures of (geographical) expertise were set up. Practically embodied in a federal insurance program, this spatial turn significantly changed the policy discourse in US flood governance, the policies and rules set out to govern the problem of floods, and the division of costs and responsibilities in this policy domain.

The case studies further show that "recognized experts" also continuously adjusted their expertise to the new political goals and values that emerged in the context of the policy field to ensure their expertise remained a valid source of knowledge in public policymaking on floods. This was so, for example, in case of the Zuiderzee Works, where engineers constantly attuned the presentation of their plan to important contextual events in the Netherlands at that time (i.e., rising unemployment, the 1916 flood, the First World War). The incorporation of ecological values by both Dutch engineers and American social geographers also exemplifies this tendency of experts to seek alignment with changes in their socio-political context, both knowingly and unknowingly. The US case further demonstrates that institutionalized expert bodies can also substantively change as a result of contextual developments. When market regulation became the cornerstone of US politics in the 1980s, social geographers were "organized out" of US flood governance, whereas the policy administrators at FEMA were "organized in".

These insights warrant a nuanced interpretation of expert-influence in Dutch and US flood governance. While experts significantly influenced the policy strategies that emerged in these policy domains, their influence was shaped by their interactions with policymakers, who drew up the boundaries within which experts could organize themselves and outline their solutions for dealing with floods. As such, the cases thus demonstrated the contextually embedded character of expert-influence.

6.4 The Impacts on Distributive Decision-Making

This section focuses on the consequences of expert-influence for the recognition and handling of the distributive implications in the decision-making process underlying the implementation of spatial measures in Dutch and US flood governance (sub-question 4).

Here, it can be concluded that the contextually embedded nature of expert-influence gave rise to close relationships between experts and policymakers in Dutch and US flood governance (see also Jasanoff 2004). Through these relationships, strong policy frames were created that, by identifying problem causes and specifying policy solutions to address these causes, constructed a coherent story around the problem of floods. In these policy frames, normative and operational arguments strongly overlapped. In the Netherlands, the institutionalized safety approach not only rested on arguments about the effectiveness of engineered flood protection in the particular context of this low-lying delta country, but also on arguments about the normative appropriateness of collective state solutions for dealing with the external risk of flooding. In the US, the institutionalized spatial planning approach not only provided a more "cost-efficient" alternative to engineered flood protection but this approach was also valued for better addressing the American values of local autonomy and individual responsibility.

However, the cases also demonstrate that contrary to general assumptions, the involvement of experts in Dutch and US flood governance did not automatically imply that the distributive implications of policy approaches were overlooked in policy- and decision-making processes. Clearly, through their close relationships with policymakers, experts greatly influenced the normative principles underlying the policy approaches that were developed in both countries. However, the cases also reveal that experts were continuously pushed by policymakers to better outline the practical (distributive) consequences of their plans.

In an effort to overcome political opposition to the Zuiderzee Works, for example, Dutch engineers and economists continuously enhanced the cost-benefit analysis that underlay this plan. This specification of costs and benefits helped Dutch decision-makers to identify potential areas of conflicting needs and interests. For example, the specification of costs for Zuiderzee fisherman who might lose their jobs as a result of the enclosure of the sea allowed politicians to raise concerns about the burden the Zuiderzee Works would put on these fishermen and their families. Concerns that were eventually—although marginally—addressed in parliamentary decision-making by granting compensation and re-education to Zuiderzee fishermen.

In the US, the interaction with social geographers provided decision-makers with information on individual insurance premiums and (the costs involved with) local damage mitigation measures the insurance program would put on home—and property owners. Based on this information, these individual costs could be traded off against other goals, such as sending effective price signals for flood damage mitigation, providing access to affordable flood insurance, and facilitating economic growth in high-risk areas. Although clearly normative and unmistakably directional, expert-influence in Dutch and US flood governance thus also provided policymakers with a blueprint for grasping and judging the distributive implications of policy measures.

This does not mean that no problems were identified with the involvement of experts in the shift to spatial measures in the Netherlands and the US. In the Netherlands, spatial measures were developed in interaction with the institutionalized body of engineering and economic expertise. In the "macro-focus" displayed by these experts, spatial measures were justified based on a trade-off between collective costs and benefits, from which it was concluded that spatial measures were cost-effective in low- to medium-risk areas. However, this macro-view falls short in fully capturing the distributive implications of spatial measures in these areas on the individual level. The extent to which and way in which spatial measures actually place costs and responsibilities on individuals in low- to medium-risk areas and how this compares to the role of individuals in high-risk areas, was not discussed in the decision-making process underlying the implementation of spatial measures in Dutch flood governance.

In the US, the situation was different. Here, the main challenge has been to uphold the interaction between spatial experts and political actors in flood governance. In the 1980s, the role of social geographers in US flood governance was significantly downplayed with the transference of management responsibilities for the National Flood Insurance Program from the federal government to the independent regulatory agency FEMA. Under FEMA's administrative guidance, the focus came to lie on restoring the operational efficiency of the insurance program on paper, pushing concerns about the program's operation in practice off the radar. The practical implications of the price incentives provided by the insurance program were no longer analyzed and communicated, and the political trade-offs that underlie the insurance program (e.g., between flood insurance affordability and providing effective price signals for flood damage mitigation) moved to the background of the policy discussion. This would not necessarily be problematic under stable contextual conditions; historical trade-offs would just remain the same. However, the US case demonstrated that climate change increased the flood risk, which, in effect, raised the costs of spatial measures for individuals in high-risk areas. These increased costs were not recognized by FEMA, because their focus was not on the effectiveness of the policy program in practice but on its sound operation on paper. When the distributive consequences of this policy manifested itself after Katrina, the trade-off had to be made in a situation of distress and unrest, where the federal government had already stepped into provide disaster relief. The description of the policymaking process after Katrina revealed the difficulty of having to reevaluate these trade-offs under tense and conflicting circumstances.

Based on these findings, it is concluded that the involvements of experts in flood governance is not principally problematic but that problems might arise from the involvement of specific types of experts in specific historical and institutional settings. Counter-intuitively, the results seem to suggest that politically-engaged experts, who are given the opportunity to successfully organize themselves in politics, can actually help to highlight the distributive underpinnings of policy choices. The analyses of the decision-making processes in Dutch and US flood governance show that because the normative assumptions underlying policy solutions are supported by expert-knowledge, they could be easily recognized and debated. At the same time, the early Zuiderzee case demonstrates the importance of having representative-democratic institutions in this regard; the Zuiderzee fishermen's opposition grew with the development of the Dutch representative democracy in the first half of the 20th century.

6.5 Main Contributions and Limitations

The cases of Dutch and US flood governance analyzed in this book show that in the shift from a safety to a spatial approach, costs and responsibilities for flood protection and damage compensation may shift from the national to the local and individual level. However, the huge difference in the extent to which and way in which local and individual responsibilities have been organized in Dutch and US flood governance—where in the US the decentralization and deregulation of responsibilities go much further than in the Netherlands—indicates that there is substantial political room to shape the distributive effects of spatial measures. To do so, it is important that these effects are recognized and discussed in the decision-making process on spatial measures in flood governance.

This book was concerned with the influence of experts on these processes. This focus originated from insights developed within the fields of science and STS studies about the political nature of expert-knowledge and its dominance in public policymaking on risks, giving experts a structural source of influence in risk governance which is not easily recognized because of its rational, consent-generating character. The conclusions of this book indicate there is a lot more to be said about how experts actually impact distributive decision-making in than is usually credited for in the academic literature on this topic. Its main contributions are twofold.

First, the findings of the two cases analyzed in this book suggest that there is nothing principally wrong with involving experts in public policymaking on floods. Based on a comparative analysis of the cases in this chapter, it was concluded that expert-influence can best be understood as contextually embedded; flood experts actively organized themselves in the policy domains of Dutch and US flood governance, but opportunities for these experts to become and stay involved in the policymaking process depended on the alignment of the knowledge of these experts with prevalent political norms and ideologies in the wider context of these policy fields. In the Netherlands, engineering knowledge fit the political ideology of the progressive-liberal elite that rose to power in the first decades of the 20th century. In the US, the knowledge of social geographers matched well with the conservative ideology that characterized US politics after Roosevelt. In both cases, close bonds were created between experts and the political elite, whose interaction created strong policy frames that presented the problem in a coherent and consistent way and in which the political values underlying policy solutions were corroborated by expert-knowledge. However, it was also found that these strong policy frames actually helped policymakers gain insight into the distributive implications of (new) flood governance solutions, which were subsequently discussed and debated in parliament. This book, therefore, concludes that the contextually embedded character of expert-influence, in some cases, actually facilitated the recognition and handling of distributive implications of flood governance strategies.

Second, this book found considerable differences between how different expert-groups impacted on the extent to which and the way in which the distributive implications of spatial measures were recognized and dealt with in Dutch and US flood governance. In particular, it found that both in the Netherlands and in the US, a lack of inclusion of "spatial-behavioral expertise" in policymaking hampered a good understanding of the distributive implications of spatial measures. This type of knowledge belongs to the disciplines of social geography, behavioral economy, and rational-institutionalism. In the Netherlands, the institutionalization of engineering expertise prevented the inclusion of spatial-behavioral expertise in flood governance, whereas in the US, this type of expertise moved to the background as a result of the outsourcing of flood management responsibilities to FEMA.

Together, these conclusions mean that rather than describing expert-influence as problematic per se, the cases warrant a more careful conclusion that looks more in particular at the specific types of expert-groups involved in specific flood governance settings. As such, it calls for research that looks beyond the role of experts as a general actor-category in public policymaking, and instead focuses on the effects of the inclusion of specific types of expert-groups in a contextualized governance setting to improve our understanding of the implications of involving experts in public policymaking institutions.

Practically, these findings suggest that spatial flood governance requires another type of expertise than standardly incorporated under a safety approach to floods. To adequately highlight the distributive implications of safety measures in the decision-making process, a new type of spatial-behavioral knowledge is needed that is able to explicate the burdens and benefits of spatial measures at the local and individual level. In this regard, this book highlights the importance of diversifying the knowledge base in flood governance when spatial measures are implemented.

This insight is relevant for other countries where spatial measures are currently being considered as a new flood governance strategy. It emphasizes the importance of incorporating spatial-behavioral expertise in the policymaking process underlying the implementation of spatial measures in flood governance. At the same time, the findings also underline the need for continuous national-level policy reflection under a spatial flood governance to be able to constantly reappraise the political trade-offs that underlie spatial policies when circumstances change, especially when they are based on regulatory instruments such as the economic incentives for flood damage mitigation provided by the US National Flood Insurance Program.

In this respect, Dutch flood governance maybe compares best to the situation in most other countries, where not so much a complete turnover to a spatial planning approach, as occurred in the US, is pursued, but rather an "integrated" governance approach is sought after, which includes both safety and spatial measures. Countries that are currently moving toward spatial measures could actually profit from their nationally-focused safety institutions by carefully planning new spatial governance institutions on top of existing institutional structures in flood governance to organize a sufficient level of national revaluation.

In terms of conceptualizations and methods, it can be concluded that the conceptualization of the shift to spatial measures as a process of gradual institutional adaptation helped with understanding the interplay of expert-influence, institutional structures, contextual factors and political agency in public policymaking. The PAF provided a useful conceptual schema to grasp and analyze this interplay. Where the PAF offered the conceptual tools to analyze to what extent and in what way policy institutions change, framing was used to understand why and how institutions adapt. In particular, this book drew on contextual and interactional accounts of framing, which highlight the interdependency between political agency, collective sense-making, and context in the framing process. This not only contributed to the recognition of the contextually embedded character of expert-influence in flood governance, but also led to a more nuanced understanding of their effects on distributive decision-making—subtleties that may be overlooked in traditional framing accounts.

However, the limitations of this research approach also should be recognized. Its main limitation lies in the case selection, and more particularly in the fact that the two "extreme" cases of the shift to spatial measures in Dutch and US flood governance were analyzed. While this helped highlight a key challenge—the integration of spatial-behavioral expertise in policymaking processes on spatial measures in flood governance—this challenge may have been magnified in these contexts. To refine the conclusions of this book, future research could examine the effects of expert-influence in more "common" cases of the shift to spatial measures in flood governance. For example, "typical" cases of integrated flood risk management could be analyzed to find out how engineering knowledge, administrative knowledge, and spatial-behavioral expertise can best be integrated into flood governance.

References

Arts B, Van Tatenhove J (2004) Policy and power: a conceptual framework between the 'old' and 'new' policy idioms. Policy Sci 37(3/4):339–356

Arts B, Leroy P, Van Tatenhove J (2006) Political modernisation and policy arrangements: a framework for understanding environmental policy change. Public Organ Rev 6(2):93–106

Bergsma E, Gupta J, Jong P (2012) Does individual responsibility increase the adaptive capacity of society? The case of local water management in the Netherlands. J Resour Conserv Recycl 64:13–22

Beveridge R (2012) Consultants, depoliticization and arena-shifting in the policy process: privatizing water in Berlin. Policy Sci 45(1):47–68

Boswell C (2009) Knowledge, legitimation and the politics of risk: the functions of research in public debates on migration. Politics Stud 57(1):165–186

Botzen WJW, van den Bergh JCJM (2008) Insurance against climate change and flooding in the Netherlands: present, future, and comparison with other countries. Risk Anal 28(2):413–426

Bouwer LM (2010) Disasters and climate change: analyses and methods for projecting future losses from extreme weather. Dissertation, VU University Amsterdam, the Netherlands

Butler C, Pidgeon N (2011) From "flood defence" to "flood risk management": exploring governance, responsibility, and blame. Environ Plann C: Gov Policy 29(3):533–547

Clemens ES, Cook JM (1999) Politics and institutionalism: explaining durability and change. Annual Review of Sociology 25(1):441–466

Dewulf A, Bouwen R (2012) Issue framing in conversations for change: discursive interaction strategies for "doing differences". J Appl Behav Sci 48(2):168–193

Dewulf A, Craps M, Dercon G (2004) How issues get framed and reframed when different communities meet: a multi-level analysis of a collaborative soil conservation initiative in the ecuadorian andes. J Community Appl Soc Psychol 14(3):177–192

Dewulf A, François G, Pahl-Wostl C, Taillieu T (2007) A framing approach to cross-disciplinary research collaboration: experiences from a large-scale research project on adaptive water management. Ecol Soc 12(2):14. Online. http://www.ecologyandsociety.org/vol12/iss2/art14/. Accessed on 24 June 2015

Dewulf A, Gray B, Putnam L, Lewicki R, Aarts MNC, Bouwen R, van Woerkum CMJ (2009) Disentangling approaches to framing in conflict and negotiation research: a meta-paradigmatic perspective. Hum Relat 62(2):155–193

Entman RM (1993) Framing: towards clarification of a fractured paradigm. J Commun 43(4):51–58

Fischer F (1995) Evaluating public policy. Nelson-Hall, Chicago

Fournier M, Larrue C, Alexander M, Hegger D, Bakker M, Pettersson M, Crabbé A, Mees H, Chorynski A (2016) Flood risk mitigation in Europe: how far away are we from the aspired forms of adaptive governance? Ecol Soc 21(4):49. https://doi.org/10.5751/ES-08991-210449

Gray B, Purdy J, Ansari J (2015) From interactions to institutions: Micro-processes of framing and mechanisms for the structuring of institutional fields. Acad Manage Rev 40(1):115–143

Habermas J (1996, translated by W. Regh) Between facts and norms—contributions to a discourse theory of law and democracy. The Massachusetts Institute of Technology Press, Cambridge

Hajer M (2005) Coalitions, practices, and meaning in environmental politics: from acid rain to bse. In: Howarth D, Torfing J (eds) Discourse theory in european politics. Palgrave Macmillan, Basingstoke, pp 297–315

Hall P (1993) Policy paradigms, social learning, and the state—the case of economic policymaking in Britain. Comp Politics 25(3):275–296

Hall P, Taylor R (1996) Political science and the three new institutionalisms. Polit Stud XLIV:936–957

Hegger DLT, Driessen PPJ, Wiering M, Van Rijswick HFMW, Kundzewiczm ZW, Matczak P, Crabbé A, Raadgever GT, Bakker MHN, Priest SJ, Larrue C, Ek K (2016) Toward more flood resilience: is a diversification of flood risk management strategies the way forward? Ecol Soc 21(4):52. https://doi.org/10.5751/ES-08854-210452

Jasanoff S (ed) (2004) States of knowledge: the co-production of science and social order. Routledge: London

Lundin M, Öberg P (2014) Expert knowledge use and deliberation in local policy making. Policy Sci 47(1):25–49

Mahoney J, Thelen K (2010) A theory of gradual institutional change. In: Mahoney J, Thelen K (eds) Explaining institutional change: ambiguity, agency and power. Cambridge University Press, New York, pp 1–37

Meijerink S, Dicke W (2008) Shifts in the public-private divide in flood management. Water Resour Dev 24(4):499–512

Neuvel JMM, Van der Knaap W (2010) A spatial planning perspective for measures concerning flood risk management. Water Resour Dev 26(2):283–296

Pahl-Wostl C (2007) Transitions towards adaptive management of water facing climate and global change. Water Resour Manage 21(1):49–62

Paudel Y, Botzen WJW, Aerts JCJH, Dijkstra TK (2015) Risk allocation in a public-private catastrophe insurance system: an actuarial analysis of deductibles, stop-loss, and premiums. J Flood Risk Manage 8(2):116–134

Penning-Rowsell EC, Pardoe J (2012) Who benefits and who loses from flood risk reduction? Environ Plann C: Gov Policy 30(3):448–466

Rein M, Schön D (1993) Reframing policy discourse. In: Fischer F, Forester J (eds) The argumentative turn in policy analysis and planning. Duke University Press, Durham, pp 145–166

Spruijt P, Knol AB, Vasileiadou E, Devilee J, Lebret E, Petersen AC (2014) Roles of scientists as policy advisers on complex issues: a literature review. Environ Sci Policy 40:16–25

Steinberg MW (1998) Tilting the frame: considerations on collective action framing from a discursive turn. Theory Soc 27(6):845–872

Swiss Re (2012) Flood—an underestimated risk: inspect, inform, insure. Swiss Reinsurance Company Ltd., Zurich. Available via Swiss Re. http://www.swissre.com/library/Flood__an_underestimated_risk_Inspect_inform_insure.html. Accessed on 3 Jun 2018

Van Hulst M, Yanow D (2016) From policy "frames" to "framing": theorizing a more dynamic, political approach. Am Rev Publ Adm 46(1):92–112

Van Tatenhove J, Arts B, Leroy P (2000) Political modernisation and the environment: the renewal of environmental policy arrangements. Kluwer Academic Publishers, Dordrecht, Boston, London, pp 53–69